Taylor & N

This is my second book & it was released in Spring 2021. I hope you enjoy the story!

Best,
Quentin Super

THE LONG ROAD EAST

Quentin Super

Copyright © 2021 Quentin Super
All rights reserved
First Edition

PAGE PUBLISHING, INC.
Conneaut Lake, PA

First originally published by Page Publishing 2021

ISBN 978-1-6624-2498-4 (hc)
ISBN 978-1-6624-2497-7 (digital)

Printed in the United States of America

I dedicate this book to my mom who, without knowing it, has blessed me with the gift of the written word. Thank you for everything.

INTRODUCTION

Revealing that I cheated on my girlfriend isn't the way to win people over. But that's what happened one November evening. I had just finished a few Grey Goose sodas with an older woman who made it clear that despite me being in a relationship, her interest in taking me back to her place was alive and well.

"Why else do you think I'm here?" she mentioned before throwing back her second shot of Patrón.

The liquor helped fight off the cold front that was taking over Minnesota as the state prepared for another winter. After finishing our drinks and splitting the bill, we got in my car and drove to her house a few minutes away.

My feet sloshed through the wet snow, quickly hitting my toes and bringing about waves of uncertainty. Feelings of nervousness and guilt swirled through my head, but still I was ready to end my relationship for the temporary pleasure that was to come.

"Take off your shoes," the woman instructed once we got inside.

We walked into her living room. Her two kids were hanging out, neither of them particularly fond of my presence. Their reactions suggested they had seen this act before.

I wanted to puke. It was pathetic that I was standing here when my girlfriend was only a short drive away.

"You guys have a good night," the woman told her kids, kissing each of them on the forehead.

She then grabbed my hand and led me up to her room. After turning on the TV, we awkwardly lied on her bed while my conscience put me in a vise grip.

Having sex with this thirty-five-year-old mother of two would not make me a better person. Doing so would brand me a cheater.

Infidelity was an act I despised, yet that couldn't repress the desire that coursed through my blood and settled into an all-familiar part of my body that had no idea how to distinguish emotion from logic.

"What do you want to watch?" the woman asked as she flipped through a few channels.

"I don't care," I responded.

The larger issue was debating with my skewed morality while in the confines of this bed.

The woman then moved closer, and soon we began making out. It wasn't long before her hands started to explore my body. I took her shirt off. She had a fantastic butterfly tattoo on her belly button. She went for my belt, and it was at that point I thought how wrong all this was.

I rolled over onto my back. I couldn't do this anymore.

"I have to go," I told her, then quickly shuffled downstairs, past her kids before hurriedly putting my shoes on and vanishing into the night.

I knew nothing would ever be the same as I drove home. I'm not the type of person who can conceal something like this. I'd rather reveal the hard truth and be hated than trapped inside the misery of knowing my brain possessed information that needed to be shared.

"We need to talk," I told my girlfriend the next morning.

That was one of the last conversations she and I would ever have.

Fast-forward to the present, six months later.

THE LONG ROAD EAST

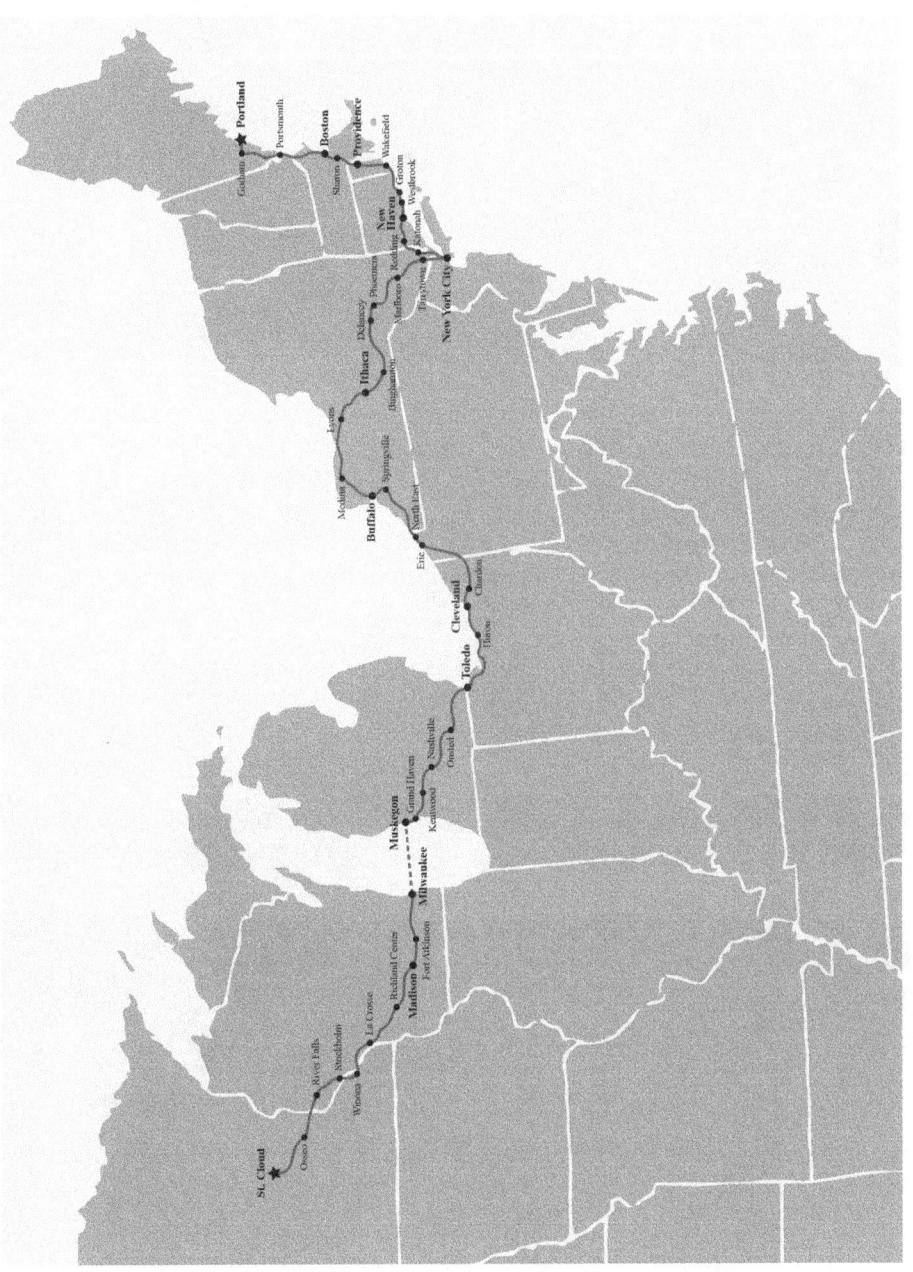

CHAPTER 1

College is finally over. It's been four long, booze-soaked years, but I'm finally getting released into the real world. To celebrate, my buddy Sam and I are jumping on our bicycles and not looking back.

We don't have a plan other than we are going to leave Minnesota and hopefully end up in Portland, Maine, before all our money has dissipated.

Originally, we were supposed to venture south to Arizona, but after realizing going there would involve sleeping under the stars, Tinder dead zones, and rattlesnakes, we wisely opted for the much more populated East Coast.

For as excited as I am to go on this trip, today I'm convinced our departure will be delayed by Sam's annoying inability to pry his bike from the clutches of a local shop.

Expecting as much, and having just returned from a sex-filled road trip, I prepare for a long day on the couch, but then at 11:00 AM my phone rings.

"Good morning," Sam says in a tone that suggests he is about to reveal something unexpected. "How are we doing today, Quen-ton?" he asks in a tone that suggests he can neither pronounce nor spell my name properly.

"I'm just lounging, brother," I inform him, all the while hoping Sam hasn't devised a plan that messes with my perfectly planned day of drinking chocolate milk and watching playoff basketball.

"I just got a call from the bike shop, and they said my stuff is ready to go. So we could ride today if you want," he casually mentions.

My blood pressure rises. My body has already melted into the sofa cushion.

"To be honest, Sam, I wasn't planning on riding today. I'm not mentally in the right frame of mind," I say.

"That's fine. I just want to ride."

I throw my shoulders back and take a deep breath. "Let me call you in a little bit," I offer, hanging up the phone and then yelling across the room at no one in particular.

"Fucking Sam," I whine as my mom walks in to check on the commotion.

"What's going on?" she asks.

"Classic Sam," I tell her. "He calls me and says let's ride. I'm not ready to go. I'm not in the frame of mind. I'm not—" I stop. "Put it this way: a lot of things are working against me."

My mom shrugs her shoulders.

"Well, isn't that going to be a lot like any other day on this trip?" she asks.

I pause and reflect on how accurate her sentiment is. It's time to swallow my pride and start our trip.

"I'll be there in an hour," I text Sam.

Minutes later my outdated bike with only three gears is loaded into the back of the family van. It's a forty-minute drive north to go meet Sam at our old apartment, enough time to have a conversation with my talkative mother.

After the minivan enters the highway and is put on cruise control, we bicker about a litany of topics, highlighted by my incessant desire to prove I know how the world works.

"I knew this woman," I begin. "Nicest woman in the world. Wouldn't hurt a fly. Grew up in a loving family. Only thing was they were highly religious. They didn't believe in sex for any purpose other than reproducing."

"Oh, here we go with the monologue," my mom sighs.

"Now just hold on a second," I say.

I then sit up in my seat as our family dog gets restless and kicks me in the balls.

"So this is the world this girl grows up in, Mom. Think about it. Small town, they don't talk about sex, and so the kids don't really know anything about it."

"Oh, this poor girl. What's your point?" my mom sighs once more.

"Well, see, the girl gets older and goes off to college. Ends up getting pregnant. The family disowns her," I add.

My mom's face isn't too concerned because she doesn't care about problems that don't belong to her. And then I continue talking, "In my opinion, that's not right. You can say the girl should have done this, or she should have done that, but she literally knew no better. And how is she going to find this stuff out? Remember, in her world, all this is extremely foreign. It's not enough to just say, 'Well, she should have just known better.' No, it doesn't work like that."

"How did we get here?" my mom asks. "Weren't we just talking about poor people or something like that?"

I reposition myself once more. My legs are getting restless, and my voice begins to crack from speaking too much.

"Yeah, I was getting there. The woman and poor people have a lot of similarities. You can't just say that someone should automatically want better if they're poor. If you know nothing other than being poor, then what's your impetus to change? Same thing with this woman. She didn't make the decision to be abstinent growing up. She was just never exposed to a world other than the one she lived in."

"If I say you're right can we stop talking about this?" she asks.

The minivan makes an exit, and it's not long before we're waiting for Sam to come out of the main entrance of his apartment building. While we patiently wait for the slow-moving Sam, my mind starts to wander. I glance left and see the sliding doors to an apartment I used to live in.

I only lived there a month, but enough debauchery occurred in that apartment to scare a small church.

Sam, a guy named Chad, and I spent that month drinking Fireball, going to bars, and doing things our sagacious fathers would wag index fingers at. I shouldn't completely implicate Chad and Sam. They were often well-behaved. The madness was mostly mine, like the time Sam came home from class one night and witnessed me involved with a Tinder prospect on his couch. "We have to think about who we are reproducing with," he scolded after my affair had ended; his words laced with so much vitriol, they were impossible to forget.

My trip down memory lane is interrupted by Sam stumbling out of the building, lifting his hand to cover the blazing sun that's darted into his eyes.

"Ah, the memories of this place. Am I right?" I ask Sam as he climbs into the backseat of the minivan.

"What are you even talking about, kid?" he asks.

Sam looks more disheveled than an addict. His words are few, and it's obvious he just wants to get on the bikes.

"I've been cleaning that damn apartment for literally the whole last week," he fumes while we make the short drive to the local bike shop.

I don't bother telling him about my last week parading around the country in a 2005 Honda Civic with a woman who needed a ride home from the Sunshine State. Now doesn't seem like the appropriate time to reminisce about my good fortune.

We pull up to the bike shop, and Sam heads inside. Knowing Sam, this ordeal could take two minutes or two hours.

"How long does it take to get a bike?" my mom asks after ten minutes.

"I think he's just having a tough day, Mom," I say.

Sam and his bike eventually come out. We're now officially resigned to spending the next weeks in close proximity.

My mom obliges my need for validation by taking a photo of us two that I can quickly throw up on Facebook.

"Is this a good photo?" she asks as she hands my phone back to me.

"It's perfect," I tell her, quickly uploading the photo before Sam asks what's taking so long.

My mom drives off and then Sam and I ride toward the Twin Cities. As soon as St. Cloud is behind us, a sigh of relief is exhaled. I don't ever have to come back here. My college town is now a safety deposit box full of sacred memories, brief successes, and hefty failures.

Twenty miles into our ride we stop at a gas station.

"I have to piss like a racehorse," I tell Sam before walking inside.

THE LONG ROAD EAST

A pang of hunger has entered my consciousness, and I take notice of the narrow aisles in this small gas station that are filled with empty calories. After using the bathroom and grabbing a Snickers bar, I walk back outside and can't seem to shake the soreness from my legs.

"Goddamn, Sam. My thighs are burning like hell," I tell him.

He rolls his eyes. "You'll feel better soon, Junior."

Still emotionally disjointed by my infidelity and unwillingness to compromise, my poor preparation for this trip is already having an effect. Instantly I regret all the weight lifting and protein shakes from the previous months. The extra two pounds of muscle added onto my Holocaust-looking frame won't help when I begin cycling through more calories than a bodybuilder.

Sam's body language is enough to know that he's unimpressed with my performance so far. As a staunch proponent of convention, he trained for this ride by actually riding his bike. That fact isn't surprising, because for Sam, abiding by convention is everything.

You will never catch Sam on the keto diet because he's not into trends. Sam is old-school. To him, anything other than a buzz-cut, Old Navy jeans and a hand-me-down T-shirt your grandpa wore when he was greasing up his Harley in 1973 is a disgrace.

While Sam's eyes slowly monitor my movements, I begin examining his bike while munching on my candy bar.

"What do you need this for?" I ask, repeatedly pointing to a number of miscellaneous objects that seem unnecessary.

"We'll see who is more prepared," Sam confidently beams as he sticks a knife big enough to kill a grizzly bear back into its holster.

"You know what I'm going to start calling you, Sam?"

"What?"

"Joad."

"Who the fuck is Joad?" he shutters.

The Joads were the family in *Grapes of Wrath*, the John Ford black-and-white film from 1940. In the film, the Joads pack up all their belongings onto this monster vehicle and head out west in search of a better life. There is a shot that shows the truck heavily weighed down by all their stuff, and it bounces up and down a dirty road, nearly on the brink of toppling over.

"You're Joad," I say. "I'm going to have to coin that and use it throughout the rest of the trip. It will be your new nickname."

"Okay, I get it. Enough with the small talk. Let's keep going," Sam urges.

We pass through a small town not too far from my parents' home. The community is quaint, and the downtown is romantic. It has a little liquor store that looks inviting. It reminds me of a different liquor store that Sam and I used to work at, the one where our boss was an angry diabetic missing half his leg.

By the time Sam and I are a few pedal strokes away from my parents' house, the sun is almost set.

"My mom has chicken wings," I mention while we blaze down a bike path with the wind at our backs.

"Nice. I'm starting to get hungry," Sam replies.

Not much has changed within my family when we arrive and take a seat in the kitchen. My hippie brother and his girlfriend are sorting through organic vegetables. In the other room my dad watches the news while my mom tries to talk to him.

"Would you be quiet?" my dad says after he's had enough of my mom's prying.

"Where are you guys headed tomorrow then?" my brother nonchalantly asks while rinsing off a vegetable I've never heard of under the cool and crisp water of the faucet.

"River Falls. It's in Wisconsin," I say.

"Oh, sweet. Yeah, that should be fun," he says, but he has no idea what the hell he is talking about because he's never ridden a bicycle farther than to the Culver's a city over.

"I think you will make it to the end of Wisconsin, and then you guys will be done," my dad says when Sam goes off for a shower.

"We'll see," I tell him.

CHAPTER 2

"That's what I like! That's what I like!" Sam sings the next morning, derisively mocking a tacky Bruno Mars song blaring from my laptop.

"You like that song, bro?"

"That might be the most annoying song I've ever heard," Sam claims.

"Sam, you wouldn't know good music if it came up and crawled in your ear," I remark.

"Enough talk now. Let's get going," he then insists.

With my dad at work, my mom is out of sorts, understandably nervous about our departure. I don't want to think about how she is feeling because even though she's my mom, this is my moment, and her emotions sully that joy.

If she could walk better, my mom would mask her nerves and go grab a coffee from Starbucks, but instead her mind is wandering without the caffeine.

"Be nice to your mom, Junior," Sam quips.

"Stop!" I hastily counter. "You know if your mom was here, you would be doing the same thing."

As Sam and I make our way outside and ready to embark, the woman who gave birth to me looks old, sixty-plus years of life inflicting wrinkles upon skin that doesn't look like it used to.

"I'm worried I might never see you again," she tells me.

She isn't wrong. I don't plan on coming back. Still drowning from a failed relationship and trying to prove I can do life on my own, living in Minnesota doesn't appeal. Right now, the rules of life seem inferior, giving me a shield against the criticism that is invited by my immature actions.

Thankfully the haze of emotional mist I used to walk around in from that failed relationship has vanished, but part of me is still missing, trapped inside the confines of an ex who was so cold the last time we hugged, it was almost as if her soul occupied a different body.

"Don't worry," I tell my mom as we prepare to ride off. "I'll be back someday."

She goes to sit on the front steps, offering her a prime view for our departure. I wave back at her and then we're off.

"Finally we get out of there," I tell Sam.

He can only shake his head at my impatience.

Before we reach the main street of the small suburb I grew up in, the trailer I'm bringing along disengages from the bar on the back of my bike.

"Dipshit, it's not on the right way," Sam notes.

We pull off to the side of the road, and I play with the apparatus connecting the trailer and bike until it properly sticks. Ergonomics has never been my specialty.

The early portion of the ride is tough. Yesterday I didn't ride with this trailer or any weight, so right away my muscles begin to stretch and pull because it's been a while since they have been called to duty. We pull up to another stoplight.

"Dude, I think I have a Joad," I whine, the strain of my trailer's weight quickly becoming an impediment to my happiness.

"And you were giving me all that shit yesterday," Sam promptly snaps.

"Yeah, I take that back. I officially have the Joad. That's what I'm going to call my trailer from now on."

Not that far from my house, we become lost. I have lived in this area all my life, but finding the street named Nashville Road is a challenge.

"This doesn't even make sense," I say after ten minutes of tapping on my GPS.

"Dude, we are on vacation. Just relax," Sam calmly advises.

"You won't be saying that if we are lost."

THE LONG ROAD EAST

Sam whips out his phone, and soon we are going the right way, but this blunder eats up two hours. We're only just outside of Minneapolis when we stop for lunch.

"This shit is kind of heavy," I quietly mumble through a Clif Bar that's sticking to my molars.

"The food?" Sam asks with a weird look.

"Um, sure," I say, not wanting to admit Joad is beyond my current level of conditioning.

"We have to get moving," Sam says.

We aren't out of the metro and already afternoon is upon us. Up hills, around side streets, in and out of bike lanes, past the University of Minnesota, and all the way across Marion Street into St. Paul we go. My back tire then hits a homeless man's shoebox.

"Sorry, sir," Sam apologizes on my behalf.

"Don't apologize! You're going to have to be more aggressive, dude," I yell back at Sam. "You are going to have to take space out here! No one is going to give it to you!"

Fuck all these people in our way, fuck Joad, and fuck Sam being nice. Two weeks ago, I was able to squat 245 pounds, but still my legs burn like an overcooked frozen pizza, the hair on my thighs pinched by biking shorts that haven't been worn in over a year.

We continuously miss turns while the GPS brings us all around Minneapolis.

"That thing is going to get us killed," I say to Sam.

For the next few hours, we battle through wrong turns and side streets, around apartment complexes, desperate for a quality route to our Warmshowers host in River Falls, Wisconsin.

Warmshowers is a website that was recommended to me by a man on a plane with more relationship troubles than me. The Warmshowers website is exclusively for people interested in the beauty of traveling by bicycle. Hosts can use the site to offer up a room in their home, a meal, a laundry machine, and almost anything that might help a biker in need. Warmshowers is also free, the idea being that people using the system abide by the pay-it-forward method.

This website will keep us out of hotels and well within our budget, but first we have to actually make it to these hosts' locations.

We are still laboring through Minneapolis, and my patience is wearing thin as the late-afternoon sun graces us with its presence. This is only day two. A continuation of cycling ineptitude will indeed leave us questioning our future at the end of Wisconsin, just like my father predicted.

Over time, we eventually grind out of the city and into the malaise of bum-fuck nowhere, also known as western Wisconsin. The sun has now set, leaving gray clouds as our backdrop for the evening. Digestion has caught up with me, and I need a bathroom break.

"We are only seven miles away," Sam complains when we stop at the next gas station.

"What do you want me to do?" I complain, wiggling my butt to get the wedgie out of my shorts. "Shit my pants?"

"Just hurry up, Q."

Sam is still perturbed when I come out.

"We can't be stopping like this, especially when we are only going eight miles an hour," he says, his words irritating and uninspiring.

"Well, what do you want me to do? I had to go, dog," I explain.

"I don't mean *now*," Sam grimaces. "I just mean later on. We can't go eight miles an hour this whole trip."

"Dude, I'm giving it all I can. I'm not cheating you on effort."

"I'm not saying you are. I'm just saying we have to pick it up."

What Sam really wants to say is that he is incensed with my abysmal speed and is questioning even doing this trip anymore. The reality is I have been giving it my all, but without acknowledging it, Sam is correct in saying something has to give. We aren't going to make it all the way to Maine with me riding like I have training wheels glued to my back tires.

Soon after the spat, we encounter a hill that isn't fun to ride up. My back squeals, and the sides of my stomach strain to pull Joad onward. It's almost 10:00 PM. Both my phone and GPS are dead, and Sam's cycling computer is also on the brink of losing functionality. The only saving grace we have is his cell phone, and the battery for that is at less than ten percent.

THE LONG ROAD EAST

I begin to worry our host has tried to make contact and, failing that, has given up on our arrival.

"The GPS says we are here," Sam claims when we turn down a sandy road, but the only thing in our line of vision is darkness and a faint streetlight.

Clueless as to what action we should take next, we creep up the road and find a gravel path with a warning sign nailed to a tree.

"Should we try it?" I ask, knowing Sam will ultimately make the call.

I can barely see my feet, and there are no lights in the far-off distance. We go halfway down the pitch-black road before Sam stops.

"I don't know, man. This is sketchy," he says. "We should turn around."

If we can't find the house, it's eight miles just to get back to town. And then we have to find a hotel. It'll be at least sixty dollars, and spending that money so soon will send me into a tailspin of self-doubt and frustration regarding my meager finances.

I call our host Mick several times off Sam's dying phone, but he doesn't answer. We ride back to a set of mailboxes in hopes of finding the correct address listed on one of them.

This time, Sam spots the house number located on the opposite side of the mailbox. But we still have to go down a different sketchy road to find the house.

Another ominous unpaved road brings us to a small cabin surrounded by trees. The lights are on, but there is no signage to indicate we have arrived at the correct address.

My damp clothes suddenly feel more wet. It's the first night, and already it feels like we are trapped inside a mystical narrative, unable to influence much of anything for these next several weeks.

"Should we try it?" I again defer.

"I don't know. It looks super sketchy."

"Let's just try. The lights are on."

Sam acquiesces to my plea. I slowly walk behind him while my teeth chatter, content to let him die first if the person inside pulls out a shotgun.

A loud bark coming from the house turns on all my survival instincts. Appearing behind the curtain of the glass door is a massive dog who shares a faint resemblance to Stephen King's Cujo.

A figure in the sliding doorway then appears.

"Hello? Who is out there?" a man asks, his right leg barely restraining the curious dog.

"Hello, sir. My name is Sam. This is my friend, Quentin Super. We are looking for Mick."

"At this hour?" he questions.

"We are bikers. We met him on Warmshowers."

"Oh, I see. Yeah, he's up the road," the man says.

"Perfect. Sorry about the trouble," Sam says, and we quickly turn around because it's any second before that man's leg gives out and Cujo comes bounding our way.

"Wait up. I can walk you up there," the man offers before we retreat. "Just let me calm down my dog."

A few minutes later the four of us are walking through slushy gravel up to the next property. The man punches in a code on a system attached to a small shed. A few noises emit from the box, and I half expect a spaceship to rise up from under the ground.

A few moments later, the rays of a flashlight appear and so does a man.

"Sorry, guys. Just saw your calls. I was sleeping. Anyway, I'm Mick," he says, extending his hand through the pouring rain.

"It's a pleasure to meet you," I tell him.

"Let's get your bikes out of the rain. Normally, I would have you camp on the grass here, but that seems like a lot to ask, given the rain."

"Any type of shelter you have is much appreciated, Mick," I say.

Sam and I stand in the storm while Mick makes space for our bikes in a shed. We are then led to a camper that runs strictly off solar energy.

"You guys drink beer?" Mick asks.

"Sam does," I say.

"Cool. Well, there are some beers and a few ice cream sandwiches in the fridge. The shower is in the back. Be conservative with it or you will run out of hot water fast," Mick says before bidding us good night.

Sam showers first. "He wasn't lying. That water is ice-cold," he says a few minutes later as he wipes down his hairy body.

I go in the back and undress. After holding my hand over the freezing water for three minutes, I defer bathing and jump straight into my sleeping bag.

A third beer having been consumed, Sam shuts off the lights and shares some more words of *encouragement* before bed.

"We're going to have to be better tomorrow," he tells me, his voice trailing off along with the night.

CHAPTER 3

"Guys, come inside. I have something to show you," Mick says the next morning.

He leads us inside his house. The interior is like a science experiment, a number of different machines scattered throughout the floors. The house itself is essentially an unfinished basement, every room visible because of uncompleted walls.

Behind Mick's computer sits a contraption he explains is representative of his belief in eco-friendly systems and appliances that use as little energy as possible.

We then go over to his computer.

"It's a trail, and it will take you the whole way there," Mick shares, diagramming a trail to Milwaukee on Google Maps.

"Are they paved?" I ask.

"Are what paved, Q?"

"The trails?"

"Not all of them. When you get to Madison and outside Milwaukee they will be though."

"Every trail in Minnesota is paved," I add.

Mick is unimpressed with my statement.

"I guess we must be spoiled over there," I say.

"If you guys don't mind," Mick says, "I'd like to ride with you for the first few miles, just to get some exercise in."

"Not a problem. Leave in fifteen?" I ask.

"Sounds good."

The sun is out as we begin riding, and already a lacquer of sweat is forming underneath my clothes. Mick and Sam jump ahead, and I'm left to stare at the muscular veins popping out the back of Mick's

THE LONG ROAD EAST

calf. For an older man, he sure isn't messing around when it comes to cycling.

The more I fall behind those two, the more likely it is Mick is asking Sam if I have what it takes. It's hard to blame him. I certainly look like a fool, carrying Joad and falling behind before any of us start to breathe heavily.

We meet up after a few miles.

"Just go slow. You don't want to pull a muscle this early," Mick tells me before heading back home.

Just like that Mick is gone, and soon Sam is way ahead of me. I finally catch up to him at a stoplight.

"Nah, you're not going slow," he assures. "I'm going about twelve miles per hour right now."

"That means I'm only going like ten, Sam."

"Ten ain't bad though."

Once we continue, bundles of energy are being expended just so my eyes can see a tiny speck of Sam in the far-off distance. It doesn't make sense why I am so far behind.

"I'm the better athlete," I murmur to myself. "How the hell is that fat boy moving so fast?"

Annoyed and vulnerable, it's too easy to blame an ancillary factor like Joad or the wind for my slow speed. I keep asking the biking gods for a break I don't yet deserve because my pride is being sapped with every mile I ride.

Somewhere in rural Wisconsin I catch up with Sam, and we stop for lunch at an outdoor restaurant that has a big chicken sitting in the front lawn. As cars pass through the drive-through, I load as many calories as possible into my stomach. Devouring this amount of food eventually makes me feel bloated, but the grease from the chicken legs has brought renewed energy to my disposition that has already been drained by the constant stress of failure.

These feelings of joy are a reminder that my pain will soon pass, or at the very least, I will learn to thrive in this state of mind. Sam is not pleased with how things are going, but he's tactfully not vomiting all his concerns on me. That's enough to make me believe everything will be just fine.

Once lunch is over, we get back to the road. After another long stretch, we stop along the Mississippi River. Sam starts taking photos, and I sit on a lonely bench and contemplate how to make my situation better.

"What's wrong?" Sam asks.

He can always tell when something isn't right.

"I don't know, man. I just have this weird feeling in my gut," I tell him.

"Like what?"

"Like things will never be the same."

"That's what it was like for me yesterday. I knew once we left St. Cloud things would never be normal again," Sam admits.

"Don't get me wrong. It's a good feeling. I just think I'm scared I won't see my family again. I'm also hesitant about these Warmshowers' spots."

The afternoon sun sparkles magnificently against the river.

"I just need to get used to living like this," I establish.

"We'll get there, man. We just need to keep the pedals moving," Sam encourages.

An hour later we arrive in Stockholm, Wisconsin, with plenty of time to spare. The directions bring us to the downtown portion of the city, but we soon realize we are nowhere near the address of our destination. Some construction workers key us in to where the address might be, so we march up a hill, only to be wrong once again. Looking to my right, there is a middle-aged woman tending to her garden.

"Excuse me, miss. How are you today?" I ask.

The woman becomes apprehensive when she looks up from her petunias. "I'm well. Can I help you?"

"We are looking for an address," I begin, showing her my phone. "But it doesn't seem to be anywhere near here. Might you know where it is?"

She looks at the screen. "I don't, unfortunately."

"You wouldn't happen to have your phone on you, would you?" I then ask. "Mine doesn't get service, and I would like to call these people we are staying with to figure out where they live."

"You can use mine," she then smiles, her heightened senses from a few seconds ago not factoring into her current generosity.

I type in the digits and then luckily am able to connect with our host Maggie.

"You're going to go up a hill to get to our place," she warns.

"No problem. There have been a lot of hills so far. We will see you soon, Maggie," I respond.

"See you guys soon."

I give the kind woman her phone back.

"Thank you for your help," I tell her.

She smiles and becomes a memory of the past.

At a T in the road, Sam and I choose to go right. The beginning of the aforementioned hill appears, but not the end. One minute into climbing the hill, my pedals are barely turning. I decide to stop pedaling and walk to the top.

"This isn't a fucking hill!" Sam crows from below. "It's a god-damn bluff!"

I look back down and laugh. This is the first time Sam has been behind me all trip. Once I finally reach the top, I look out and see nothing but an expanse of farmland.

"She said it was only two blocks away once we got up the hill," I tell Sam once he catches up.

"Well, what the hell?" he muses in frustration.

"I don't think Maggie knows what two blocks look like, bro."

Visits to three different homes atop the grassy plain yield no results. I continue to grow more frustrated. We arrived in good time, but now we are spending over an hour trying to find a house in a proximity *maybe* fifty people live in. My blood boils while thinking of Maggie's words, then wondering if she is even a real person or a donut-filled slob behind a computer in Delaware.

We stop at a fourth house and run into a younger woman who fortunately is familiar with Maggie and her family.

"They live off that county highway," the woman says, pointing to a map on Sam's phone.

Begrudgingly, we backtrack to the other side of the hill, but the address soon comes into view. We get off our bikes and begin walking forward.

A German shepherd charges from my left before skidding to a stop ten feet away.

"Jackson, get over here!" a man screams at the protective dog.

"How are we doing, sir? You must be Jim," I say, offering my hand.

"Jim's up there," he says without shaking my hand, so we continue walking.

Two lesser imposing dogs then come bounding toward us, along with a young boy who has no underwear on. His little *thing* is gyrating, and when Maggie soon appears with her child in her arms, it doesn't take long to learn that her family lives differently than most.

"Hello, hello. Glad you guys found the place okay," Maggie smiles.

She is so short her eyes meet my belly button. Based on the way her nipples poke through her tank top, she isn't wearing a bra. In noticing the number of animals running freely around her property, one might think we went the wrong way and are in South Dakota.

"Welcome to our home," Maggie says, then leading me downstairs while Sam unpacks his essentials inside a barn.

"I hope you don't mind that someone else slept here last night," she tells me from the steps of her chilly basement. "I haven't gotten a chance to wash the sheets yet either."

She then points to a mattress on the floor.

"Not a problem," I tell her, noticing a couch off to the side. "Sam can have the bed."

After showering, we all congregate around a campfire for dinner. Halfway through, disaster almost strikes when one of the kids nearly falls into the firepit.

My body seizes while watching Maggie make a desperate lunge for her offspring. She grabs the child before a nightmare has been lived out, but still the event has put everyone on edge.

"That was close," I note.

THE LONG ROAD EAST

Maggie's husband Jim tries to make peace with the situation, but there is no denying the tension the near catastrophe has brought.

"So how did you guys meet?" I ask when it feels like an appropriate amount of time has passed.

"In college," says Maggie. "We had known each other since we went to the same Bible camp way back in elementary school."

"That's far back."

"We were both homeschooled."

"Interesting," I softly acknowledge.

The few homeschooled people I know are strange, in line with all the stereotypes you hear about children who learn math from their mom with a degree in third-wave feminism. This fact also might help explain why Maggie is nonchalant about breastfeeding her child as we all relax after dinner.

Before the campfire songs come out, I excuse myself and head downstairs to scroll through Facebook, and then a half hour later Sam approaches.

"Do you want some eggnog?" he asks. "They're making some upstairs."

"No. I'm good, bro," I reply.

"It's fresh from their cows though."

Agony rumbles in my stomach. "That just doesn't sound enticing, dude."

I lie on the couch for the next hour talking with Anne, the owner of that 2005 Honda Civic from a few days prior.

"I miss you," she says at one point in our conversation, which is irritating because I'm gone and not coming back anytime soon.

"That's so nice of you to say," I manage to dole out.

"We should be a couple," she suggests.

"Not a good time. You have to give me space, Anne, otherwise you are going to push me away."

"I'm just worried you will never talk to me again," she says.

"I'm on a bike tour. I literally ride my bike all day. I'm not ignoring you. I'm just out here doing shit, you know?"

"I know, but it seems like you don't want to talk to me."

"No, that's not the case," I lie.

Talking with Anne is not a priority, but it's nothing to do with her. It's just difficult to look backward when each day brings me further away from my old life.

I manage to make a few promises that are easy to keep, and Anne finally hangs up the phone. I breathe a sigh of relief. For now, at least one problem has been avoided.

CHAPTER 4

"My stomach doesn't feel so good," Sam utters the next morning. "I think it was the eggnog."

"No shit it was the eggnog," I laugh while imagining unpasteurized milk particles floating inside Sam's flummoxed intestines.

"I think I need a minute," he grunts.

"Bro, you're going to need a half hour. Why would you drink raw eggnog?"

"Raw eggnog? That's not a thing."

"Whatever it is. You deserve to shit your brains out."

"Shut the hell up," Sam beams.

Once Sam's stomach has adjusted, we then begin another day. That morning we pass small towns along the Mississippi River that serve as tourist stops during the summer. Sam is snapping plenty of photos, just like Rodney Dangerfield's Asian friend in *Caddyshack*, making sure no one confuses us with being locals.

"I have to eat," I say at one point, interrupting Sam's next *National Geographic* cover shoot.

"There is probably a gas station up ahead," he says.

"It's risky to just assume that. This town isn't that big. I don't want to ride hungry."

Sam looks back at the Kwik Trip that is one hundred yards behind us.

"I don't like to backtrack," he says.

"I don't either, but the Kwik Trip is right *there*," I emphasize, my eyes widening while pointing at the building containing donuts and chocolate milk.

"You and eating, Q," Sam smirks.

Acquiescing to Sam's desire, we proceed forward without sustenance, operating under the belief that another gas station is not far. Road signs say the next city is eight miles ahead, but when we arrive, the place is deemed unincorporated.

As an aside, if anyone tells you to visit an unincorporated city, don't go, because there isn't anything there.

My saving grace comes eight miles later in a small town just a few miles outside Winona, Minnesota.

"Can you make it to Winona?" Sam asks when we stop at a Kwik Trip.

"Why don't we just stop here?" I complain with the force of a whining toddler.

"I don't want to make two stops so close together," he says.

"I need to eat, dog."

"Fine. Just hurry up."

After walking inside, a testy cashier tangles with my path. "Sir, you can't park in the handicap spot," she says.

"I didn't drive here," I tell her.

"Yes, but your bikes are in the handicapped spot."

I want to tell her it'll only take a minute for me to be in and out, but she's standing with her hands on her hips, right in front of the chocolate milk door.

"I'll go tell my friend we have to move," I cede.

"Thank you, sir."

I walk back outside. "Sam, we have to move the bikes."

"Why?"

"They can't be in the handicapped spot."

"Jesus Christ."

Five minutes later I come back out with egg sandwiches and a bottle of chocolate milk. A few employees are sitting on a picnic table smoking cigarettes and talking about the weather. I turn my head toward Sam.

"Aren't you glad we don't work at the liquor store anymore?" I joke, but he's in no mood to reminisce about the past.

"Hurry up and eat that. We're almost there," he says.

Once we reach Winona, we are supposed to meet a woman named Pat, but after ringing her doorbell for the third time, it's clear she's elsewhere.

"Let's go get something to eat," Sam insists.

My head aches. "We just ate, dude. And we just got here."

"She's not here though."

"Yeah, but we don't need to be in a rush, man."

"I'm just going to leave then," Sam insists.

"And go where?"

"Somewhere. I don't know."

"Sam, what the hell? I don't understand. What's the problem?"

"I'm not just going to sit here!" he belts, blowing up like a purple balloon.

A response to his dissatisfaction is difficult to conjure as we silently sit in the front yard for a few moments. It's a privilege to be here with Sam, but he's masking a frustration toward me that has been brewing ever since I got us lost on the second day of the trip.

A young girl from the neighboring house peers through a curtain as we continue to sit patiently in the front yard. I look at Sam for a reset, but his mind and eyes are elsewhere. When I look back the other way, a graying woman is creeping around the corner.

I stand up and flash her a smile, but she is looking at me like she has no idea who the hell is standing in her front yard.

"Pat?" I ask.

"Yes. I'm Pat."

"I'm Quentin Super. And this is my friend Sam. We had talked a few days ago on Warmshowers about us staying here tonight."

"Oh gosh, I'm sorry. I was at yoga. I didn't think you guys were coming after you didn't respond to my message," she says.

I think back to our exchange. Being the irresponsible prick that I am, I never did send her the all-cementing confirmation message.

"That's my fault, Pat. I should have confirmed with you after we made the initial plans. Is it still okay if we stay here?" I ask.

She smiles and then motions for us to come inside.

After soaking in the shower for twenty minutes, I sit down next to Sam at the kitchen table. "Look, man, I'm not expecting an apology—" I say.

"That's good, because you're not going to get one," he interrupts.

I shake my head. "All I'm saying is what happened, happened. We can either let this fester or move on."

Sam pauses, not ready to accept one iota of blame, but altogether realizing that holding resentment over feelings in Winona won't benefit us if a crisis emerges in the middle of Ohio. He squints his eyes and nods his head, and I interpret that to mean a resolution has been reached.

After dinner, the three of us sit down to watch *Witness*, a Harrison Ford film from before my time.

"It's a commentary on temptation, and what kind of people resist or don't resist," Pat says at the conclusion, referring to a scene in the film where Harrison Ford's character chooses not to sleep with an Amish woman.

Guilt floods into my brain. It wasn't just the one woman from the bar that put me in this position. Tinder's got me in a chokehold.

CHAPTER 5

Today is only a twenty-three-mile ride. This evening we will be staying with Sam's cousin in La Crosse, Wisconsin. Today is also the barometer for how the trip is going. If after arriving in La Crosse we are relatively unscathed, we will simply continue on as is, but if Sam decides he is carrying too much weight or I feel inclined to buy more equipment, then this stop will be a prime opportunity to recalibrate.

"Do you want to get on the trail?" Sam asks underneath a shining sun.

"I thought it started in La Crosse."

"My GPS is saying we can hop on it soon."

"Do you think it will be paved?" I ask, thinking back to the conversation with Mick.

"I don't know. It might be."

"I don't care, I suppose. I mean, it's only twenty-something miles," I figure.

As soon as we enter the trail, I freak.

"It's all gravel!" I yell.

"Crushed limestone," Sam calmly corrects.

"Either way, fuck me. This shit is horrible."

"I'm going to ride ahead. I'll stop in a little bit," Sam assures.

I only get a few miles into the trail and already my composure is being tested. While passing a marsh in the middle of a forest, I look down to see a snake positioned a few feet from my front tire. A sharp rush of fear takes over, and my hands shake. The snake hisses, and it feels like forever waiting for it to bite me, but it never does.

"Jesus fucking Christ!" I wail in despair.

I'm probably going to die on this trip, and that's fine because Sam can write my eulogy and tell everyone I died in the name of Tinder and crushed limestone.

After catching my breath, I stop on a bridge above a small creek to check how far away the destination is. Once the coordinates are plugged in, my eyes can't believe what they are reading. Time to call Sam.

"Hello?" says a voice in a tone that is none too distraught.

"Yes, um, Sam, have you looked at your GPS lately?" I say in trying to be civil.

"No. Why?"

"Well, we are going *way* out of the way on this trail. We have added about ten miles to our day."

"Excellent," is his disenchanting response that leads me to stumble and look at my phone to make sure the right number was dialed.

After a few seconds of silence, I hang up.

Excellent, according to Sam. Morning sex after a night of drinking is excellent, but in no way is adding miles to our already-humid day excellent.

I ride further before deciding to use my other unreliable piece of technology to get me to La Crosse via an actual road. The GPS takes me off the trail, but quickly a big hill comes into view. While riding up, another pest enters the fray.

"Good afternoon!" says a jolly biker who scares the hell out of me. He says a few more words before speeding ahead.

"Fuck you, man," I whisper under my breath, questioning everything at this point.

Notions of independence begin to manifest. I can keep going without Sam, not just to La Crosse, but to Maine. Today feels like a betrayal, that Sam is purposely taking the trail to make me suffer.

I soon come to a stoplight in the busiest part of the afternoon. I look to my right and serendipitously see Sam perched on a cement overlook taking photos of the sun bouncing off the river. I exit the road and park fifteen feet behind him. He keeps checking the trail and then his phone, clueless to my position.

I don't say anything, choosing instead to let him writhe in the unknown. I'm a bitter, ungrateful, disrespectful jerk, but carrying all those labels is better than telling Sam I bailed on the trail to save myself from the agony of snakes and crushed limestone.

A few minutes pass before Sam conveniently turns around and sees my lurking presence.

"How long have you been here?" he questions.

"You didn't see me?" I taunt, feigning confusion.

"No."

"I biked right past you. I was gassed, so I just got to the top of the hill."

"Really?" Sam asks, believing my lies. "I thought you got on the road or something."

"Nah," I say, chuckling with a devious grin.

A confession would be appropriate, but there are more important matters at hand.

"So, look, we are going to have to have a conversation about this trail," I say and then adjust my pants so my thighs chafe slightly less.

"What do you mean?"

"I don't think riding on the trail is going to work out."

"Why not?" Sam asks, resounding disappointment attached to his diction.

"I'm getting killed on there. I feel like I'm going to pop a tire. And"—I pause to lend credence to my next words—"I saw a snake."

"But the trail takes us all the way to Milwaukee," Sam replies.

"Yeah, but, bro, it's not working for me. And truthfully, I think we should do what's best for the both of us. Not just you."

Sam's upset.

"What I mean is that you are going to ride well no matter which route we take. For me, I know I'm not going to ride well on the gravel," I tell him.

"Crushed limestone," Sam heeds once more.

"Right. Crushed limestone," I reiterate, making air quotes for emphasis. "Anyway, so this crushed limestone is a burden. I think it boils down to simple math. Either we have one person thrive on the limestone, or we have two people thrive on the road."

"You're not thriving on the road," Sam reminds.

"Well, I think I am, and that's important."

"But what about the tunnels?"

"What tunnels?"

"Remember, the tunnels throughout. They're supposed to be really cool."

As I prepare to go into a discourse on making sacrifices in life, Sam's eyes have the look of a kid who really wants the happy meal at McDonald's. In the moment, I can't break his beating heart, no matter how little the stupid tunnels appeal.

"Let's just talk about it later," I say, confident I will get my way eventually.

To get to Sam's cousin's home, we go through downtown La Crosse at the busiest hour of the day. Construction is taking place along the main drag, so there is only a one-lane road with no shoulder.

I race ahead, but when I turn back, Sam is not in sight. I pull off to the side and wait for one minute, hopeful that he will tiptoe into view.

No luck. Time to take out the phone.

"Dude, I got turned around. I'm going to send you the address," Sam says when I reach him.

My phone's battery is on life support. Trying to use the GPS will surely kill the battery. I take a screenshot of the directions and simply hope for the best.

Fortunately, I know La Crosse fairly well. It's the same city where Sam told me to not base a good time on getting laid. It's also where one night I witnessed people flipping over cars and Sam walking around with bloody toes hanging out of his sandals.

"La Crosse is where your liver goes to die," some guy told me when I first visited.

He wasn't wrong.

After pulling up to the same Kwik Trip where we used to get food the mornings after to soak up all the alcohol, there is no doubt I'm in La Crosse, yet the dangers of having a dying phone still exist.

"Don't die on me, you stupid fucking thing," I keep chattering to the inanimate object, periodically checking my location.

By the grace of the biking gods, and with two percent battery remaining, I pull up to the address just as Sam turns the corner. His cousin Beth greets us in the front lawn and then shows us to the garage.

"You can put your bikes in here," she invites, then gives Sam a hug.

"What did you go to school for?" she asks me a few sentences later.

"I have a master's degree in English," I say.

"Cool. I'm working on mine right now," she responds.

"Nice. Yeah, I feel so thankful to have one at my age."

"That's great," Beth sneers. "Where are you guys going tomorrow then?"

My head rears toward Sam. We're supposed to be staying here two nights, but I guess Sam never mentioned that part to Beth.

"Sam, do you have something to say?" I push.

"I was actually going to talk to you about that, Beth," he starts, the redness in his face visible even through his scruffy beard. "Maybe we could stay another night, if you guys are cool with it."

Beth smiles. "Of course you can," she says.

Soon, Beth's husband Matt comes home from work with a few pizzas in tow. "Q, do you want something to drink?" he asks.

Everyone else has cracked open a beer, and being opposed to barley and calories, I need something more potent to pique my interest. "To be honest, I'm more of a wine and Grey Goose guy, Matt."

"Oh, I have wine," he replies, then leading me to a small fridge with an array of flavors. "White or red?"

"White, please."

He hands me a glass and then the bottle.

"I can drink the whole bottle?" I excitedly ask.

"Uh, I guess, if you really want to do that."

After a few slices of pizza, the wine has already begun to go down dangerously quick.

My friend Lacy arrives around this time, followed by more members of Sam's family. As the Smiths talk about family matters, I make sure to polish off the entire bottle of wine. Boredom and libido draw my eyes to the other end of the table where Lacy patiently sits.

It's not known where she and I stand. I tried seducing her a few years earlier, but back then she was into a guy who rode motorcycles. Tonight is different though, and it will not end before I make a pass at her.

"It sounds like we're going to go downtown for a bit," Sam tells Beth as dinner concludes.

"Have fun," she tells all of us as we then scurry out the door.

Already feeling frisky from the wine, there is a level of excitement to the next few hours. The first bar we visit is a recreation of those 1930s drinking spots where people wore suits and drank dirty martinis. There are a few blue suede couches off to the side, as well as dark, elegant wallpaper that gives the place a tone of sincerity.

"How are we doing tonight?" I ask the bartender with a black shirt and ponytail. "Can I get a Long Island?"

I forget to tell him to leave out the tequila. That libation and I have a jaded relationship because every time I indulge, the next day is spent with my head in the toilet.

Sam and I had talked earlier about how we wanted to celebrate our first few legs of the journey. We're certainly doing just that, but after a bottle of wine and a Long Island, I am beyond celebrating. It's now approaching recklessness.

"Do you guys want to go home?" Lacy asks as we soon exit the upscale bar.

Sam and I look at each other. "Definitely not," we say collectively.

Going home would be a wise choice, but I still want to pursue Lacy. From that point, the night gets blurrier. At the next place we prepare to do a Rudy Bomb.

"Is there tequila in this?" I ask Lacy.

"Yeah, why? Are you allergic?"

"Basically."

"Well then don't drink it," she says, reaching for the shot.

"No, no. I'm going to take it," I tell her, then throwing back the nastiest concoction my taste buds have ever experienced.

I saunter over to the near-empty dance floor and begin gyrating out of rhythm. A few minutes later, Sam and Lacy join, but a few minutes after that, Lacy is screaming into my ear. "We're going to the next spot!"

"Okay, fine," I mumble.

The next place is packed, and maneuvering through the crowd presents a unique challenge. "I got this round!" I declare once we find some open space, and then Sam and Lacy tell me what they want.

After shimmying up to the bar, I'm greeted by an attractive woman with a face that couldn't be more disinterested in my presence.

"What do you want?" she yells over the music.

Suddenly unable to remember my friends' drink orders, panic ensues.

"I want a Grey Goose Red Bull," I yell back.

The bartender begins to walk away.

"Wait," I yell after her. "I also need a whisky Coke and a tequila sunrise."

Those drinks are not what Sam and Lacy want, but Sam will drink anything and believing Lacy likes tequila isn't exactly foolish. I hand my credit card over to the snobby woman and after signing the check bring the drinks back to my friends.

"What's this?" Sam asks.

"It's what you ordered," I drunkenly chuckle.

"No, it's not."

"Yeah, it is. You said whiskey Coke, right?"

"No, but whatever," he says.

"I definitely didn't order this," Lacy says, holding her drink with a limp wrist.

"Fuck, well, the bartender looked new," I chuckle, and then go back to dancing like a buffoon among the mass of people.

The three of us try to dance in a crowd that has no room for my lanky frame; after a few songs my head begins to beg for water.

"We should get him out of here," I hear Lacy tell Sam.

"That's a good idea," he says.

My legs are so sore I can barely stand.

"Come here, big guy," Sam instructs, and then he and Lacy carry me out of the bar draped over their shoulders.

The last thing I remember is falling into Lacy's couch, but what I *don't* remember and have to be told the next day is sloppily propositioning Lacy and sticking my odorous feet in Sam's face.

Sadly, I do remember waking up the next morning, rushing to the bathroom, and green goo violently exiting every orifice of my body.

CHAPTER 6

After an hour spent puking my brains out in the bathroom, I meander back to the comforts of Lacy's couch. It's there that I roll around until my head stops spinning just enough so I can hear Lacy tell me Sam left in the middle of the night because he could not tolerate the putrid smell of my size 12 feet.

"Sam also tried to sleep in my bed," Lacy informs.

"That's not shocking," I respond.

"But when I said no, he took an Uber home."

Lacy then begins to search the apartment for a missing piece to her wardrobe. She is dressed like a giant cat.

My head still spinning, I can't figure out why.

"I have an event today," she says.

"An event? You look like you're going to a Halloween party," my distressed brain conjures.

"I'm a mascot."

The thought of wearing a costume tickles my gag reflex. This is one of those mornings where part of me wishes last night didn't happen, that all the alcohol in my bloodstream could magically dissipate.

"You can drive, right?" Lacy asks as she hurriedly gathers more essentials from the messy floor.

My head hurts, and I don't understand why humans choose to drink alcohol. Life would be simpler right now if I had chosen to drink responsibly.

"Drive? Maybe," I reply.

My head is rattling with more force than that snake by the marsh. It takes a Herculean effort, but a few minutes later I'm able to get off the couch and trudge my way into the escalator and down the stairs toward a car that will jumble the pile of debris waiting to exit my stomach.

Lacy drives us to her school. Getting behind the wheel at this moment may be a criminal offense. It all depends on if I'm hungover or still drunk, but expressing my concerns would prove futile because Lacy is in a hurry, and my present condition is the furthest thing from her mind.

"Drive my car back to your place and I'll get it later," she instructs, closing the door and then running through the gymnasium doors.

The keys are still in the ignition. I keep shaking my head, disgusted and ashamed this is the situation I've put myself in. I move over to the driver's seat and let out a burp. It would be a travesty if something horrible happened in the next twenty minutes.

I flick on the radio in desperate search of pop or rock music, but the only thing playing is a country singer still drowning in his own sorrows.

After putting the car in reverse, my head begins to jolt with annoyance. I can physically drive, but my head is so temporarily concussed that having to debate the morality of my actions is not practical.

After making a few turns, a new problem surfaces. Whatever is in my stomach now needs to come out. It's hard to maintain concentration after this happens, and I'm praying to whatever deity I don't believe in that my sphincter doesn't lose grip and make a mockery of Lacy's front seat. This sequence is like the countdown to a rocket launch, each upcoming turn the car makes adding more discomfort to my repressed bowels.

Before making a mess, I finally arrive at the house. The front door is locked so I waddle around to the back, convinced I am a few seconds from uncorking in the yard. Thankfully, the back door is open.

Ten minutes later I can take a deep breath, but still every part of my body hurts. My legs are sore, and I'd need to drink at least ten glasses of water to flush out all the tequila still coursing through my blood.

Experience tells me I'll feel better by the end of the day, but right now it's barely noon. The path to feeling better has become a waiting game, so I go upstairs and fall onto the bed.

When my eyes crack open a few hours later, I summon the courage to walk downstairs and search for food. It's been over twelve hours since I last ate.

"Where's Lacy?" I ask a spry and refreshed Sam.

"She came for her car and then left," he explains.

"Fuck. I was hoping to talk to her."

I settle on a couch and stare at the blue wall. "Bro, I feel like shit," I tell Sam.

"You know how these things go, Q. It'll pass."

Thanks to Don Julio, this hangover will not soon pass. I try eating a banana, but my stomach rejects its advances.

"I'm going to puke," I tell Sam before sprinting out the back door.

Sam has a concerned look on his face when I return. "Where did you throw up?" he asks.

"By that tree in the backyard," I murmur, plopping back down on the couch.

"Q!" Sam snaps.

I look at him, unsure of his gripe. "What?"

"The dogs run around out there. They can't eat that."

"Sam, they're not going to eat it. Relax."

"You don't know that. That stuff has to be cleaned up."

I begin violently moving around on the couch, my way of begging not to be forced to go back out and make amends.

"Just sit there," Sam instructs. "I'll go fix this."

This is hangover purgatory. I can't eat, but my body needs sustenance to feel better. And now I'm probably on the hook for doggie homicide by hangover.

As the day turns to late afternoon, I still feel like trash. Sam's phone dings.

"Who's that?" I ask.

"Beth."

"What does she want?"

"She wants to know if we want to go pick mushrooms with her and Matt."

This hangover suddenly doesn't seem so bad, because even if I were as sober as a nun, there is no way I'm going to go pick mushrooms.

My parents used to drag my brothers and I out to this "marble mound" that my uncle owned. In reality, it was this huge pile of dirt that contained one ugly, broken marble for every thousand feet of useless waste. My parents would get so excited to go, like we were going to Disney World or some other place that had a pool.

"I feel sick," I'd say, trying to force them to leave me behind.

"Quent, we might go to McDonald's if you behave," my mom would say, knowing full well that restaurant violated the stringent health codes my parents so judiciously sought to enforce.

Sam stands up to go to the kitchen, and I'm still rolling around on the couch like a whiny child who's being forced to go to bed at 9:00 PM.

"Just think about it," he says while rummaging through the fridge.

Matt and Beth soon come home. Sam is jovial and ready to go, but I continue to wallow on the sofa.

"So, Q, do you think you're up for picking mushrooms?" Matt asks after changing out of his work clothes.

I sink deeper into the couch. "Actually, do you guys have any Pepto-Bismol?" I ask.

"He's not feeling too well," Sam notes.

Beth doesn't like where this is going, but she's too nice not to share the pink, day-saving medicine. "I have some," she huffs.

The three of them soon leave. After chugging what's left of the Pepto-Bismol, it only takes thirty minutes before the previous night's sins are washed away. Feeling like the luckiest guy this side of the Mississippi, I stroll down the block and get my haircut by a round mound with purple hair. She doesn't believe biking to Maine is part of my future plans.

"Uh-huh," she sarcastically nods as I tell her of the itinerary, and she must be slow in the noodle because it should occur to her that I can see her looking cross-eyed at me in the mirror directly in front of us.

After tipping 10 percent and brushing a few strands of hair off my shirt, I venture over to the grocery store that's only a few steps away. Some of the food I throw into my cart is for the road, but most of it is for tonight. The prize of it all is a rotisserie chicken that I nibble on for the remainder of the evening while watching *Forrest Gump*.

I'm no stranger to this movie, but on this occasion, seeing Forrest's journey, hearing "Running on Empty" by the Jackson Browne band, and Jenny eventually succumbing to the wrath of AIDS breaks me.

I sit alone on the couch for a couple minutes happily crying, thankful for the life that has been given to me.

CHAPTER 7

Matt told us how to get out of La Crosse, but somehow we are going the wrong direction again. Sam's a people pleaser, so we aren't taking the snake-infested trails and are instead going to trek across Wisconsin using highways and traditional roads.

In this case, the only person Sam is pleasing is me. He's doing so because to do otherwise would involve too much time spent having a debate that I intellectualize in hopes of outsmarting him to get my way.

Perhaps fittingly, the hills are the worst they've been all trip. Some stretch vertically over a mile long. The hangover is over, but the feeling of having no control still remains.

When we stop at Kwik Trip to scarf down corn dogs, Sam isn't smiling. Seeing him without a smile is rare. My best guess is he too hates the difficulty of the roads. I don't tell him to get his mind right because going up towering hills makes me resent much of present reality as well.

A normal hill takes a few minutes to conquer, but these ones draping across the open heartland of Wisconsin are mountainous and backbreaking. A stop in pedaling could easily see one of us rolling backward.

"We have to keep grinding," I tell Sam.

While slowly creeping up another hill, a semi passes on my left. It sounds like a dinosaur being wheeled to its death. The hinges on the wheels ache with despair. The flashers strum like a heartbeat, slow and aware of every ounce of liquid running through their arteries. My friend Dan would call these Arby's hills because they're so tall that most people would give up, roll back down, and hope a drive-through appears.

For Sam and I, this is just another obstacle that has to be overcome.

Carrying weight up these winding stretches leaves my back squealing. By the time I get to the top of each hill, stable breathing eludes my lungs. I can't look forward to some big payoff at the top either. Coasting down the other side only means that the beginning of a new ascent awaits.

"Atta boy!" I yell to Sam at the top of one hill. He has pulled over to the side and is taking more photos. The eighty-year-old rendition of himself will be thankful for all the memories he is capturing.

All day I haven't been able to embrace the moment in the same way Sam can. At times it's possible to appreciate putting my kickstand down or hearing the faint clicks of Sam's yellow bags when he snaps them shut, but most of the time reality feels like a drag, the only cure being to fantasize about the women I may encounter either at a bar or on the Internet.

With sunlight becoming more rare, it's only eight more miles until we meet another new face, courtesy of Warmshowers. At the top of another backbreaking hill a few miles outside of town, I realize Sam has gone too far. When I tell him as much over the phone, he is none too pleased.

"I'm sorry, man. I hate to tell you to turn around," I say.

"It's whatever. I'm coming back."

While waiting for him to rejoin, I begin trying to contact our Warmshowers host. She hasn't answered her phone all day, and I'm growing more nervous with each call that goes unreturned. Further worry mounts when Sam begins closing in.

I call the woman once more.

No response.

I call her again…

No response.

Sam arrives and then I conveniently get a text.

"I'm sorry, but I can't host you anymore," the message reads.

"I don't understand," I write back. "It's late afternoon, and we won't be able to find another Warmshowers in time. You're really letting us down."

"There's a campsite a few miles away," is all she can muster for a reply.

I turn to Sam. "This fucking chick," I cry.

"What happened?"

"She's bailing on us."

"Why?" he asks.

"I don't know. I should bash her on Warmshowers. Save someone else this hassle."

"*Don't* go doing that," Sam wails.

"Why not? She fucked us over."

"It's bad karma. It might come back to bite us later if we talk bad about her now."

"Then what do we do?"

"Junior, I think you and I both know what has to be done," Sam confidently says.

He's right. It's time to bite the bullet and pay for a hotel.

"This is my bad, bro. I got the hotel," I tell him.

Sam has already trekked the few miles back, and now he is going to have to ride back into town. "Dude, don't worry about it," he says, which is surprising to hear because I expected another verbal lashing.

We laugh and trade stories about how our problems don't matter because we are doing what some people only dream of. I live for these moments with Sam, the ones where we are on the same page and everything feels so right.

In the months leading up to the trip they became less and less because my impulses had a stranglehold on my priorities. And Sam spent every day thinking of something new that could go wrong, proceeding to then go on Amazon and throw money at a problem that didn't yet exist.

Our friendship lost its identity during this rough patch. We initially bonded over booze, racquetball, and brotherhood, but as university neared the end, so too did our chemistry.

"I don't think we should live together after the trip," he said at one point, citing the floundering relationships each of us had with people we once loved and then despised after living together.

He wasn't wrong, but I didn't want him to be right. I would miss looking to my right and seeing him sitting on his shitty aqua leather chair, drinking boxed red wine from a glass that maybe saw the dishwasher two weeks ago.

There would be no more playing video games, ruthlessly screaming at each other over a meaningless virtual reality. There would also be no "you want to go downtown tonight?" inquiries, which even as going to the row of college bars became less fun, the invitation itself was still warming.

Sam and I came up together during university, and while we still had a long way to go in terms of figuring it all out, I felt privileged to be able to go through a number of formative years with him.

I type in Richland Center on the Kayak app's search bar. Every hotel is around one hundred dollars except for one place that says to call about the price.

"You said the bed's broken?" I confirm with an older woman at the hotel.

"No," she stresses. "The mattress is on the floor, so that's why we are offering a discount."

"The mattress is on the floor? That sounds broken. Can I get the room for sixty dollars?"

"No. I can only give you 10 percent off," she claims.

"I'll go elsewhere then," I say, hoping she comes down.

"Okay. Bye," she says, then hangs up the phone.

"Damn it," I say to Sam, shedding the phone from my ear.

"What happened?"

"Long story, but I think the chick on the phone was a bit touched."

We get into town and I examine Kayak one more time, and still none of the places are offering a good deal. To our right is the hotel I just called. It even has a restaurant.

"Fuck it. I'm gonna try this place again. Wait here," I tell Sam before walking inside.

The woman at the front desk doesn't recognize my voice. The price for the room has not changed. After deciding it's best to rent a room, the woman hands me two key cards.

After Sam and I take turns showering, we walk downstairs to patronize the restaurant.

"You were only supposed to go up once," Sam informs me when I come back to our table with a fourth bowl of salad.

In looking at the menu, the unlimited salad bar the waitress sold us is actually extremely limited. "Seems a bit strange to call it unlimited then," I snap.

"I'm sure it will be fine," Sam says. "I'm almost done. Let's pay and get out of here."

I go back to the room and call Anne before falling asleep.

"I miss you," she again tells me.

"Anne, I'm on this trip," I say resentfully.

"When can I see you?" she asks.

"That's tough. Realistically, the only way would be is if you drove out here and got a hotel room. Then Sam and I could spend the night there."

Sam won't want to spend a night in the same room as Anne and I, but he will be interested in a free hotel.

"Do you have a friend that could come with?" I ask Anne.

"Maybe Steve."

"The fuck? I was thinking female, Anne. You know, maybe someone for Sam?"

"I don't think so. Not on short notice. I have an idea though."

She tells me the plan.

"You're going to have to drive a long way then," I inform her.

"It'll be worth it."

With Anne, it was always worth it.

CHAPTER 8

The roads continue to wind up and down as we make our way to Madison, Wisconsin. Each day thus far has consisted of sixty-mile rides, but every day recovering from these difficult pulls becomes surprisingly easier.

At this point in the trip, we know what to expect. I know what time Sam will tie his shoes every morning, when we will stop for lunch, and when my legs will tire underneath the afternoon sun.

It's day 8 of the trip, and we have ridden for seven of them, and nearly every day we have added miles to our itinerary because of errors both in and out of our control. That has to change as the trip progresses because the road will likely only wear us down further.

Over the course of the next few days, the plan is to go from Richland Center, to Madison, to Fort Atkinson, and then end Wisconsin in Milwaukee. Our only relatively easy day will be from Madison to Fort Atkinson, and even that will be around forty miles. This is what we expected and what we are prepared to do, but there is a huge difference between talking about something and actually doing it.

Later on, twenty miles into the ride, the headstrong wind has derailed my psyche. Part of me wants to quit because the constant hills and winds have made riding unenjoyable. It's also difficult to find motivation right now, my brain circuitously wrapped around my infidelity and replaying the multitude of decisions I could have made differently.

When I later see Sam's bike, it means a chance to take a break. He is sitting outside, using two hands to shovel food into his mouth under the awning of a burger joint. A warm fuzzy feeling forms in my throat knowing I can let the grease of a burger simmer in my trachea before washing it down with a creamy milkshake.

"Sit down," Sam instructs as I labor through dismounting my bike, my asshole feeling like it is going to rip apart. "I bought you a meal."

"You what?" I smile.

"Just sit down, kid."

Elation kicks in. I could give Sam a kiss. Out comes the server with a burger and fries, along with the coveted strawberry milkshake. The bill is thirty dollars, but it feels like Sam has just bought me a new car.

"Do you want some money?" I ask when the last drop of milkshake makes its way up the colorful straw and into my joyful mouth.

"No. It's a gift. Let's get back to it. We still have a long way to go," a recharged Sam says.

The meal was great, but the road continues to impose its will. The combination of hills, miles, and beaming sun has erased all the satisfaction that was felt from the late lunch. I reach into my pocket and grab a pair of headphones. Music will be my escape through these last miles. A few songs I've heard too many times come funneling into my eardrums, but then on comes a Five Finger Death Punch track called "Nobody Praying for Me."

It's been three years since this song sounded good, but right now it vibes with my current disposition. The loneliness I've felt from not only this trip, but also from the last five months hits deeper as the lyrics blend together. I was supposed to feel better about my relationship failures as time has elapsed, but instead I seem to only feel more conflicted with who I am as a person.

We turn off a busy road and onto a slower one filled with lakeside homes. My mind flirts with the thought of staying in a million-dollar mansion, but as we get closer to the destination, the homes gradually lessen in economic value.

It is at our next Warmshowers that we meet Chelsea and Abby. They're two lesbians staying with the hotter one's parents.

"Madison is like a vacuum," the mom explains during dinner. "Once you get outside the city, it is very Republican."

"Could you further elaborate?" I ask, knowing full well the red-based Sam will not like the temperature of the room.

"Certainly. Milwaukee and Madison are very progressive cities, but virtually the rest of Wisconsin is conservative. If you were to look at a color-coded map of the state, only us and Milwaukee would be blue."

"That's fairly typical though. They're both bigger cities," I say.

"Yes, but enough about politics," the mom then says. "You guys want a beer or something?"

"I'll have a beer with you," Sam replies.

The thought of asking her if she has wine crosses my mind, but then memories of La Crosse spring up, and the notion is quickly retracted.

The entire meal the dad silently lurks on the far end of the table. He doesn't have much to bring to the conversation. He is more there as an observer, perhaps curious to see if I will make an advance toward his homosexual daughter.

"Did you guys ever go to college?" he asks.

"Yeah, I just got my master's degree," I tell him.

"Oh, I see," he says, unimpressed with my accomplishment.

The old man doesn't find my presence likeable, and it's hard to tell if it's me or if he doesn't like any stranger bearing a penis. This continues even after we all move into the living room to relax after the long meal.

"You feeling okay there, Junior?" Sam asks after thirty minutes of small talk has floated into the atmosphere.

"I think I'm going to go to bed," I announce, having never felt there was anything I could bring to this surface-level conversation.

The dad and his unbecoming gaze follow my feet as I agonizingly remove myself from the leather sofa cushions and head toward the bedroom.

"You just want to go on Tinder," Sam jeers while I limp away.

With a wink I say good night and stumble away.

"You guys have to be out by ten tomorrow," the dad says before I reach the door.

"Yes, sir. Thank you for your hospitality," I solemnly mutter.

CHAPTER 9

The next morning, I wake up to a message on Tinder from a woman named Amy.

"Hey! How are you?" her template-like message reads.

"You know I'm from out of town, right?" I write back.

"I figured," she responds.

"I suppose I have to ask, what are you looking for?"

"Whatever. I'm down to fuck, or even a relationship," she claims.

"Well, I can't do a relationship, but I can definitely have sex with you. Only problem is I'm heading to Fort Atkinson today. Thoughts?"

"Come over during my lunch hour," she says back, but that will never work with my schedule.

When I ask Sam for his opinion, he is not keen on the idea. "I'll ride ahead, and you can catch up later. I'm not waiting around for this *maybe* chick," he says with disapproval.

The more I remain in contact with Amy, the more time slips away from our morning. We are going to have to leave soon.

"Meet me in Fort Atkinson," I tell Amy.

"I'm down for whatever," she says. "I'd rather get to ride you once than never at all."

And with that there is suddenly a newfound urgency to get to Fort Atkinson. By the time we leave, everyone except one of the lesbians has vacated the premises. We go over to a gas station for breakfast, and Sam says what I don't have the balls to admit.

"That sucks Chelsea is a lesbian. She's so hot," he states.

I almost keel over in laughter. "You can't say things like that, dude."

"Why?"

"The social justice warriors would have your head."

"You don't agree?" he asks.

"No, I agree. Chelsea is sexy as fuck, but that doesn't mean she has to be straight," I say, much to the chagrin of Sam's wrinkled nose.

The ride to Fort Atkinson is cake, but when we arrive at Jim's house in the middle of the afternoon, the sky is overcast, and it looks like it might rain soon.

"You guys want a beer?" Jim asks after we shake hands.

"I'll be honest, Jim. Nearly everyone on this trip has offered us a beer. Not a bad thing. Just interesting," I say.

He gives me this look like, "I didn't ask for your life story. I asked if you wanted a beer."

"I know you guys are from Minnesota, but in Wisconsin, we drink beer for virtually every social event," Jim continues.

Sam and Jim enjoy their beverages while the news plays on a TV hanging from the wall. Pest, Jim's dog, an annoying animal that can't stop giving me death stares, begins barking whenever my foot moves. The dog is clearly not privy to strangers, and his barks gradually become nastier. Then Jim's daughter passes through the living room on her way upstairs.

"Stay and chat, hun," Jim says.

"I can't. I have homework to do," she tells him.

"Oh, come on," Sam pipes in with a smile.

I shoot him a glance that intimates, *Please don't try to fuck the farmer's daughter, Sam.*

"What's up?" Sam asks upon noticing my concern.

"Nothing," I tell him.

"But you just looked at me strange."

"Just calm down. Don't go getting the wrong idea, Sam."

"I think it's you who has the wrong idea."

"Just drink your beer, bro," I say, and thus ends any worry I have about Sam pulling a questionable stunt.

Jim begins talking, but the only part that makes sense is that he likes to ride his bike long distances. Unlike us, whenever he rides, he is always on a time schedule.

"I really admire what you guys are doing. You don't get these kinds of opportunities all the time," he says.

"Thank you. We're grateful to do it," I tell him.

"You guys hungry?" he asks. "We're having dinner later tonight."

"I'm hungry right now. I was thinking Sam and I could walk into town and get some food."

I turn to Sam. "Are you hungry?"

"I could eat," he admits.

"No problem. Why don't we all convene here later then," suggests Jim.

"Sounds good."

Sam and I venture to a local Mexican restaurant, and when the chips and salsa come out, I inform him of my evening plans with Amy.

"I don't know, man. It sounds risky," he says.

"What's risky?" I ask.

"You trying to bring a girl into Jim's house."

"Dude, I'm not going to bring a woman into his house. That dog would never let that happen," I tease.

"I wasn't going to say anything, but it bit my foot," Sam utters.

"We should kill it, tonight," I chuckle only half jokingly.

"Where would you go then?" is Sam's legitimate question when we come back to the subject of Amy.

"Probably her car."

"Jesus!" he wails.

"Oh, come on. What?" I ask just as the server brings out the main courses.

"It's just—" Sam begins, but then I cut him off.

"Hold on. She's calling me."

A few minutes later, it's now confirmed Amy is coming. All I have to do is relax and enjoy the rest of the evening, provided Jim's rabid dog doesn't pull a wild stunt that impales my plans. We finish our meal and then head home.

"How are you planning on doing this?" Sam asks as we amble on a sidewalk through a middle-class neighborhood.

"Doing what?" I ask.

"This Tinder thing."

"Oh, that," I say. "I'm just going to say I have a friend in the area that wants to meet up. We are going to go out for ice cream, and that I'll be back in a short while."

"I don't know, man. That sounds risky."

"It's not like that, Sam. And stop saying risky. You're not a CEO."

When we return to the house, Pest is still doing everything he can to make sure my foot ends up in his face, and the family has sat down for dinner.

"You guys want to go for a walk after dinner?" the wife asks us.

"That sounds fun," I lie, but I need to be seen as likable for when I disappear later.

"We can go get ice cream after," the wife then smiles.

She's too happy for a cynic like myself, and her desire for an ice cream social puts a dent in my plans. It's time to scramble for a new excuse as to why I will be leaving their home later that night.

After dinner, the family changes into athletic gear, and we drive to the park. As soon as I shut the car door, fifty-six mosquitoes glom onto my body.

"Why didn't you wear pants?" Jim asks me, and I just smile because if he knew why I was really here he would throw my ass out on the street.

He leads us into the tall grass as every patch of exposed skin on my body continues to get sucked dry by the mosquitoes.

"Be on the lookout for ticks," Jim cautions.

I look down at my shorts and flip-flops. They won't do a damn thing if a Lyme-infested creature attaches to my lower body. Jim then begins rambling about nature, further widening a gap between our universes. Frustration mounts because my mind isn't interested in plants. It's centered on Amy and her desire to ride me endlessly.

By the time we mercifully get back to the car, it's dark out, and Amy hasn't yet given me an arrival time.

"We still good?" I text her.

Thirty minutes later and still no response.

At the ice cream shop, Jim and his wife are behaving like newlyweds, smiling and laughing at every joke they share, even sneaking in kisses when no one is looking.

"Is this 1965 or 2017?" I ask Sam.

"I don't understand," he says.

"Jim, and his wife. Do you not see... Ah, fuck, forget it, man."

I take my phone out of my pocket and find the screen empty. Getting laid seems improbable at this point, and I've just wasted three hours fantasizing about having sex in a car.

It's 10:30 PM when we get back to the house. I sneak off to call Amy in a last-ditch effort, but she doesn't answer. For the 594th time in my life, rejection has occurred.

While dejectedly climbing onto a pullout mattress, my phone buzzes. It's a message from Amy.

"OMG, I'm so sorry. My dad had a heart attack," the message reads.

CHAPTER 10

"Fuck that chick," I tell Sam as we prepare to make our way to Milwaukee.

He breaks into a fit of joyous laughter. "What's wrong now, Q?"

"The chick from last night said her dad had a heart attack. She's just lying because she wanted to bail. Man, that shit is so annoying."

"Jesus, kid. What if her dad actually had a heart attack?" Sam asks, still rollicking with unabated cackling.

"Well," I begin, a smile creeping across my face after having embraced the vanishing of a golden opportunity. "Then fuck her dad."

"You're a savage," Sam chirps.

We leave Fort Atkinson bright and early, prepared to get to Milwaukee at a decent hour. The plan is for Sam and I to meet with Anne and her friend Tim at a motel in the city. How this situation will play out is unclear, but my motivations are simple: I want to see Anne.

The probability of getting laid tonight is high, which gives me an extra gear burning through the pedal strokes.

"This pavement sure is nice," I tell Sam, feeling confident as the miles tick away fairly easily.

Neither Sam nor I confirm we are going the right direction, but we are on a trail, enough justification to ignore any lingering doubts.

Per usual, Sam has pulled ahead, but a few miles later he begins riding back toward me. He certainly isn't coming back to share how much he admires my uncanny ability to consume chocolate milk.

I stop pedaling as his approach slows. We slowly converge and come to a simultaneous halt. His eyes never meet mine.

"I got some bad news," he says, and then comes the pause. "We're going the wrong way."

I look into the far-off distance. "That's what I figured."

My feet shuffle to mitigate the tension and frustration. Once again, we can't get out of our own way.

"How far do you think we have to backtrack?" I ask.

"I'm thinking all the way," Sam replies.

"All the way, like all the way back to the next trail crossing?"

"No," Sam quietly answers. "Like all the way back to Fort Atkinson."

"Fuck." I take a deep breath. "Well, let's get after it. Nothing we can do but ride."

Simple as that. There is no time to mope and point fingers. We just have to be better.

The GPS fortunately doesn't take us the entire way back to Fort Atkinson. Sam exits the trail at a highway overpass, and we are soon bound for the campus of UW–Whitewater.

A slew of emotions, mostly negative, continue to circulate my brain. I don't want to do eighty miles after expecting to do only sixty. I also don't want to jeopardize a surefire good time with Anne, but the reality is I haven't taken enough responsibility to ensure that our rides go smoothly. All I've done is assumed and expected Sam has every route mapped out, but that mindset is silly because this is unchartered territory for the both of us.

When we reconnect at a gas station to take a break, Sam looks at me in a way only he can. His smile radiates and brings a new energy to the situation.

"How are we feeling?" he asks, his jovial tone enough to brighten even the darkest of days.

A smile beams from my lips. "Getting there, bro. We're really hammering out these miles."

"Fuck yeah we are. We will be just fine."

"I am going to treat myself to a slushie," I say before walking inside.

I grab a slice of pizza and then carry it over to the slushie machine and grab a small cup. The blue candy that flows from the machine is so enticing, I slurp down half of it on my way to the cash register.

THE LONG ROAD EAST

"That'll be $6.23," the cashier tells me, her eyes indicating she would rather be anywhere but behind this counter.

Her body language makes it easy to be appreciative of just where it is we are. Sam and I are utterly failing in many aspects of this trip, but we are also out here together, following our dreams and grinding through whatever adversity is presented. Even if today we tack on another forty miles through sheer stupidity, there is still nothing else I would rather be doing.

"We're not as far off as I initially thought," Sam notes as a painful freeze begins to take hold of the veins in my brain. "We'll be alright if we keep a steady pace."

"Let's just keep grinding, bro."

I throw my slushie into the garbage bin, and we continue the long trek toward Milwaukee. When we stop at another gas station twenty miles outside our destination, the sun hasn't even begun to set.

"Here, take this extra ice," Sam says, holding out a clear plastic bag that he just purchased.

I dump as much ice as possible into my CamelBak, the same one that has served as a water bottle throughout this entire trip. After a few swigs of old, warm water, out comes a refreshing stream that scintillates my dried-out taste buds.

"I like the way we're riding," I tell Sam.

It's palpable how locked in we both are. We both want to atone for the earlier blunder and finish this day strong.

As we resume riding, a decrepit road, uglier than a gap-toothed cat, enters the fray. Thud, followed by thud, followed by thud. Every five feet, divots in the pavement cripple my lower back. I stop for a moment just to make sure I'm still on an actual road.

"Jesus Christ," I tell myself. "This is terrible."

Toward the end of this frenzied stretch, I see that up ahead Sam is waiting at a stoplight. The red light is turning green, and as I begin to push forward, my pedals no longer are rotating. In a panic, I slam my feet against the pedals in attempt to push them through, but they don't budge.

I raise my hand to get Sam's attention. This is our first mechanical issue, and like any time this was bound to happen, the timing is inopportune.

Sam raises his hand in response, interpreting my plea for help as a signal to go ahead. "Oh, goddamn it. No. Sam!" I yell, but he can't hear me. He steps back on his bike and rides through the intersection just before the light changes. This forces me to fight off soreness and run after him while dragging my bike and Joad along a tight and narrow road.

There is no way I'm going to catch him, but there's too much adrenaline pumping to think more rationally. I get to the corner and whip out my phone, desperate for signal and battery.

"Pick up, pick up," I say, and then he luckily does.

"Just go through the light," are Sam's first words.

"Dude, you have to come back. My pedals won't turn."

"What do you mean they won't turn?"

"I don't know, but I can't ride my bike because the pedals literally won't move."

There is a pause. "Fuck, give me a few minutes. I have to turn around."

Sam returns after a few minutes and without hesitation goes to work on my pedals. His hands soon become covered in grease, miles of gravel and pavement now rubbing onto his sweaty palms. When he stands and wipes his hands on a towel, my pedal problem is eradicated. They make a circular motion once more, and just like when he bought me a burger and milkshake, I cannot thank Sam enough.

"We just have to keep moving," he says. "I can take another look at it when we get to the motel."

We ride parallel the entire way into Milwaukee. I want to work my ass off as payment for his selfless effort. Momentum has been regained, and all seems perfect until my pedal problem recurs once more a few miles from the motel.

"Don't shift the rest of the day," Sam instructs while fidgeting with the chain once again.

"Like at all?"

"Yeah. Don't shift gears. Something is going on with the gears."

THE LONG ROAD EAST

This second deterrence means I'm not dealing with a small problem. We have to find a bike shop sooner than later.

As we finish the ride, I am burning all my energy, tactfully pushing harder when we climb upward and then conserving stamina on flats and declines. It's hard to resist the urge to gear out at various times, but staying in neutral is better than having to deal with the chain detaching.

When we pull into the motel parking lot that sits directly parallel with Highway 94, the sight of cars blazing down the highway eighty miles per hour is not lost on me. Worries about my bike are postponed because we have made it this far. Unlike my father predicted, this is not the end for us. It's only the beginning.

***

Anne and her friend Tim meet us outside the motel. The building has an eerie feel to it, like one of those motels you see in movies where people do drugs and kill each other. Patrons drinking out of skinny brown paper bags sit on balconies and stare down at us with suspect glances. Nothing about this place is inviting, not even the janky pool set up inside the perimeter.

"You look really tan," Anne tells me, distracting me from concern.

"Duh. We've been in the sun for over a week."

Anne is wearing a black tank top, a colorful skirt, and some unflattering red boots. Something about the way she looks bothers me, like she didn't put enough effort into her appearance.

The four of us move to the back of the motel where a tired, beaten-down woman and her three rambunctious kids gawk at our bikes. After seeing we have nothing to say, the kids begin running around the parking lot, an action their mother does not appreciate.

"Don't ever have kids," she mumbles to us.

Tim comes back with two sets of key cards. "One for us, and one for you guys," he says, handing me one of the white key cards.

The entrance to the building is impeded by cracked concrete and yellow tape that urges readers to be cautious. I unhook Joad and lift my bike over the tape. Anne is calmly standing outside next to Joad when I return, so I sensually run my fingers up and down her surgically repaired spine.

"It's nice to see you," I tell her.

She doesn't say anything, instead smiling in a flirtatious manner so as not to give too much emotion away. When we walk up to the room, Sam and Tim are engaged in a conversation about the sleeping arrangements.

"This whole setup doesn't work for me. It's too awkward," Tim observes.

"I can't sleep on this floor," Sam then says, pushing his foot against the hard surface.

"Why don't we just figure it out later?" I interject. "Is anyone else hungry because I could eat the ass end of a boar pig."

"But what about the bed situation?" Tim asks.

"There are two beds and four people. We can easily make this work, but right now we should go, otherwise I'm going to get irritated. I need to eat."

Reluctantly, Anne drives us just outside the city limits, where we settle on an Italian restaurant. The server is polite, calling everyone sir or ma'am at least twenty times. Table talk is reserved to Anne and Tim annoyingly laughing at everything that's been said.

Sam is clearly annoyed. I'm embarrassed because he deserves better than a surface-level conversation with two immature teenagers. If I had half an ounce of self-respect, I'd tell Anne and Tim to leave, but I'm a slave to my sexuality.

All four of us know what tonight is about, but to explicitly acknowledge that truth and act accordingly would be considered socially incorrect.

When the check comes, Anne graciously reaches for it. "I got it," she says.

"No, let's split it," I hastily reply.

"I want to pay for it," she demands, brushing my hand away.

I'm not in a position to reject charity, but Anne did already drive eight hours from Minneapolis to get here.

"I don't know what to say, but I do really appreciate the gesture," I tell Anne.

Anne signs the credit card slip and then we all shimmy out of the booth and leave the restaurant.

"Can we stop and get some beer?" Sam asks.

"Of course," Anne tells him.

As we drive to a gas station, I open Tinder to see what kind of women live in eastern Wisconsin. After a few swipes, Sam is giving me a dirty look. He shakes his head and turns away to look out the window. Shrugging my shoulders is the only response I can muster.

Taking the time to explain my relationship with Anne to Sam would prove futile. He's not like me when it comes to women.

We finally make it back to the hotel after an hour spent driving around looking for a place that sells booze. After a few sips of Mike's Hard Lemonade, I set the glass bottle on the nightstand and patiently wait to see how the night will play out.

For the next hour, tedious small talk pervades the room while everyone negotiates space and nonverbally claims sleeping quarters. Anne and I are lying together on one bed, trying not to be too touchy for fear that Sam or Tim might explode and demand a change.

As a sports update makes its way into the latter half of the news program on TV, I look over and see that Tim is asleep. Not the "I'm going to close my eyes and hope I fall asleep" kind of slumber either. It's stage 2 of the REM cycle.

I keep lifting my head over Anne's shoulder to see what Sam is doing. He's not within eyesight, but I keep listening for his clumpy gait to indicate where he is. The more I crane my neck around Anne, the more I'm convinced Sam is not here, but I can't place when he left the room.

Anne keeps looking at me with excitement.

"I'm going to check the bathroom," I quietly tell her while getting up, still not sure where Sam could possibly be.

The bathroom door is ajar, and the light is off. I should be more concerned with my best friend's whereabouts, but so long as he isn't being mauled outside by that weird mom from earlier, all is well.

"Come here," I whisper to Anne, motioning with my hands so as not to wake Tim.

I point at an open space next to the sink. "Sit here," I command.

We begin having sex. I feel emotionally connected throughout, so wrapped up in the moment that I want to tell Anne I love her even though I don't. The blue in my corneas is glistening from the light bulbs emitting waves just above the mirror. Right then every miscue of the trip seems necessary, part of a plan to bring me to this very moment.

Afterward I wash my hands and walk back to the bed. Sam is still gone, and Tim is still sleeping. I lay my head on the pillow, and Anne curls up next to me. Eyes fluttering, my phone dings. A text from Sam.

"Went and got my own room. Not a big deal. See you tomorrow," his text reads.

I figure my next visual will be sunlight breaking through the curtains, but around 3:00 AM a warm pair of lips move slowly across my stomach.

No words need to be said. I put on another condom, and Anne rides me for the next ten minutes. I keep looking over at the other bed, sure Tim will break out of his slumber and see the two of us. His face is turned in our direction, and it keeps throwing off my concentration.

Every few minutes Tim readjusts, but his face stays the same. I keep thinking his eyes will randomly open and catch mine, and then we will share this awkward moment.

Not that it matters. Life is great right now. Eventually the ending comes, that five-second interval right before a full-blown orgasm. I bury my head in Anne's shoulder to smother the ecstasy. She rolls over, and we both laugh. This is what we do.

CHAPTER 11

"See you later," I tell Anne the next morning.

"I'll call you when I get back to Minneapolis," she says with a smile.

She's now going to catch more feelings and begin inundating me with Facebook messages, constant pleas for me to give more effort toward a relationship we aren't in. She knows there is nothing inside my crushed soul that can be shared with her, that my heart still belongs to someone else, but the entire time I've known that a part of her has felt she could change that.

The worst part of this is she's going to hate me once she realizes that a future between us will never happen. She's going to think all guys are like me and then carry that resentment with her for the rest of her life. Hopefully she knows I care about her, but if she knew how that alone isn't enough, I'd instantly become a pariah in her eyes.

Her and Tim walk out of the room, and I grab for my phone.

"You ready?" I text Sam.

"Meet you downstairs in five," he says.

We still have to find the ferry that brings us over Lake Michigan before it leaves in two hours. We are only five miles away from the lake, but with each passing minute there becomes a bigger chance for problems to arise.

It wouldn't be a catastrophe if we miss the ferry. It simply would mean we would have to spend another day in Milwaukee, but this adventure is about moving forward, not staying in place.

When we convene outside, Sam explains his absence.

"I just bought another room," he tells me. "But then the key card didn't work so I had to call the front desk."

"Seems normal," I say.

"Not at all. The guy who walked up had a gun sticking out of his belt."

"Like in a holster on his hip?" I ask.

"No, Q. It was in his belt. That shit was so ghetto."

I nod my head, not really knowing what to say.

"I could just sense what was happening," Sam then segues.

"What do you mean?"

"You two were getting all comfy and Tim was already asleep. There was no way I was sleeping with him."

"We could have made other arrangements. You didn't have to buy another room."

"Dude, spare me the bullshit. You didn't give two shits where I was last night."

"That's not entirely true, Sam. I was confused though because I didn't know where you were."

"Whatever. We don't need to talk about it anymore," Sam says authoritatively.

We ride in silence to the docking port. When we get there, all of our stuff is searched by a man surprised to see two cyclists wanting to take the ferry.

"Long ride today?" he asked.

"Not at all," I told him.

"Where are you going?"

"Portland, Maine," I inform him.

"Ah, well, good luck."

We are let through and then greeted by an old man working at the front desk. He offers first-class accommodations for a nominal fee.

"Are you interested in first-class, bro?" I ask Sam.

"Sure," he grumbles.

"We'll take the first-class," I tell the man, and as he punches in some information on his keyboard, I look back at Sam to see if he's still cranky.

"Here's two tickets. Enjoy your trip," the man then says, handing over two thick white pieces of paper.

I tap Sam's shoulder, and we go outside to sit on a bench and wait for the ferry to arrive. As I munch on a few crackers, Sam goes over to chat with two middle-aged Germans who are also biking around America.

"I wish I would have done that when I was your guys' age," I can hear the German man tell Sam.

"We are really lucky, sir," Sam responds.

Soon the ferry arrives, and people from Michigan get off and walk away from the vessel. We are slowly let on, with Sam, myself, and the two Germans walking our bikes onto the main deck, where they will need to be secured against the wall.

One worker hands me a red strap, and before we make eye contact he walks away. I'm now holding the strap with a look of defeat, another inanimate object having imposed its will on my brain.

Trying not to look completely incompetent, I wage war with the strap until it loosely secures my bike.

"You have to secure that better," a man tells me as the intercom above announces that the ferry is about to leave.

"Could you help me?" I ask.

His look insinuates no one has ever asked him that question.

"I need help," I repeat.

"Okay, fine. Just get up to the decks," he tells me.

I trudge up to the first-class area. One positive to upgrading our tickets is that the next two hours will be spent away from little kids running amok and begging their parents for candy. In the first-class cabin there are only businessmen and an Asian guy walking around and taking people's orders.

Sam is in the corner. I take a seat right next to him, his expression livelier than it was even fifteen minutes prior. As I get comfy, he reveals a large smile and motions to the Nikon camera hanging around his neck.

"Just so you know, I've already been to the deck, and it's really cool," Sam informs me.

"Really? It looks cold outside," I say, basing my analysis off the wind that is throwing people around like rag dolls just outside the window.

Soon into the voyage, my stomach is wrenching. The small bumps from breaks in the waves are accentuating the ringing in my head. The entire time I do nothing but look out the window because the Wi-Fi has proven to be awful, and this doesn't feel like the upgrade promised by that old man back on land.

Sam begins pacing around the rig with a cheerful smile. He's holding his camera with two hands and soon becomes entrenched in conversation with random strangers. It's likely he and whomever he speaks with will bond over surface-level topics that would make me exit the conversation, only to get chastised later for doing just that.

"We will be arriving in Muskegon in a few minutes, folks," the captain announces.

I stumble up to the top deck to look at the lake that stretches for miles. On the shore, two grungy dudes are spitting into the lake.

"There is a lot of crime in Muskegon," a person from Wisconsin warned. "You don't want to be there when it gets dark."

My dad always cautioned against similar danger, convinced the darkness covered up all the criminals who otherwise would be seen as normal people during the day. I am not dumb enough to test the Wisconsinite's theory, but it's hard to believe nighttime turns the city of Muskegon into Gotham.

Once the ferry docks, I make my way down to my bike. The employee who helped fasten my bike to the wall definitely did his job. In fact, now I can't get my bike free from the clutches of the straps. People and cars are beginning to file off the ferry, leaving only me and the German woman to struggle helplessly.

She sees my angst and comes over.

"Can you help me with my bike?" she asks.

"I don't know how," I respond, and then she saunters dejectedly back to her post.

The other employees, all men it should be noted, are busy directing cars off the ferry. They are quite impressive, all decked out in dark blue outfits. Too shy to ask for more help, I hope one reads my mind and comes over.

On the other side of the deck, Sam is casually walking his bike off the ferry. I contemplate yelling for help, but that would only anger him.

A young man goes over to help the German woman and then comes over to me once I get his attention. "Do you need any help?" he asks.

"Of course. I can't get my bike unhooked," I tell him.

He helps and then I exit the ferry.

"We don't have too far to go," I tell a disinterested Sam while looking at my GPS in the parking lot.

"Good," he brims.

The western Michigan sky is littered with overcast clouds. They mesh well with the light green leaves attached to trees. If it rains the moment will be lost, but for now I'm invited to witness the small lakes and cool humidity that will make this ride peaceful.

Snapping a photo with my eyes, this is a level of beauty I won't soon forget, but trying to capture this moment is futile. I will never be able to recreate this feeling.

With five miles to go before we reach our Warmshowers, the battery in my phone is almost drained. I take out my other GPS, but it soon has us going the wrong way, so we pull over into a near-empty parking lot. A man outside a row of small businesses comes out from his shop to ask if we need help.

"You're from Minnesota?" he soon asks.

"Yes. We are trying to go all the way to Portland, Maine," I inform him.

He smiles. "I'm jealous, guys. Do it while you're young."

The directions we received aren't the best, and in no mood for further extracurriculars, Sam reclaims his navigating duties. A few turns later we arrive at another Warmshowers. A younger man comes out of his front door enthusiastically clapping. It's the nicest reception we have received all trip.

"What's up, guys!" exclaims the man.

"Hey, hey! How are we doing?" Sam exudes with similar gusto.

"Welcome to my pad, guys. I'm Rob," the man announces.

We take a quick tour around his place. One interesting item is a large school bus that has been renovated for family vacations. It has a bed, kitchen, and dinner table.

"You guys don't have to sleep in here," Rob says. "There is also the couch or my daughter's room. She's going to be at her mom's place tonight."

"We'll take whatever," I laugh.

We then move to the kitchen to acquaint ourselves more properly. "I've noticed that people on Warmshowers are a little more put together," says Rob, noting the difference in people who stay at his house compared to other sites like Couchsurfing.

"We take a lot of pride in that," I respond.

"I want my kids to meet all different kinds of people," Rob goes on to explain. "Of course, that has its drawbacks. One time a guy stayed here for a whole week because his wheel was busted. Finally, I was just like dude, here, take this wheel, and you gotta fucking go. But in general, people are pretty good."

"I agree. People have been *very* hospitable thus far," I say. "Have you used Warmshowers yourself?"

"Not really, but I took my wife and kids to New York last summer. Drove our bus out there as a family vacation. We used Couchsurfing, and the people let me park on their property. They just did their thing, and we stayed out of their way. Couldn't have been happier."

"It seems like only nice people host on Warmshowers, but I'm sure there will be some people on this trip that will not be as accommodating," I say, more as a warning to myself than anything.

"You'll have to let me know," Rob then laughs.

For dinner, Sam and I go to a Mexican restaurant to watch the first game of an NBA playoff matchup between the Cavs and Celtics. When we come back, Rob is sitting in the living room.

"Cavs are absolutely dominating tonight," I tell Rob.

"Were you guys watching the game?" Rob asks.

"Yeah. It was on TV at the restaurant."

"Wow, come have a seat. If I knew you guys were into ball, I would have gone to dinner with you," Rob says as Sam and I melt into the sofa cushions.

Suddenly, two young women walk in from the kitchen.

"Ah, boys. This is my sister-in-law Jamie and her friend Abby," Rob says.

Sam and I shoot out of our seats to shake their hands. My disheveled appearance won't win over any hearts, but I try to play it off by puffing out my bird chest a little more.

The two women's glances and flips of the hair intimate they are at least moderately interested. This evening suddenly feels like an opportunity.

"What are you guys doing tonight?" Rob asks the women.

"Going to Slicks," answers Jamie.

"Is that a bar?" I ask.

"Yeah," says Jamie. "It's got music and dancing. Cheap drinks too."

"Why don't you bring Sam and Q?" Rob suggests.

"That'd be cool," Jamie nonchalantly says.

My mind races with possibilities. Perhaps women from Michigan aren't as conservative as those from Minnesota and Wisconsin. If a quarter century under the guise of traditional values has taught me anything about the feminine psyche, it's that I should always prepare for the unexpected.

"Get dressed and we'll meet you guys outside," Jamie instructs.

A few minutes later, Sam comes out from the bathroom.

"Bro, you're dressed all fancy," I joke.

"What do you mean?" he asks with a sheepish grin.

Every day on this trip, Sam has put on a black button-up and khaki shorts after showering. The outfit looks nice on him, but by the third day seeing him in those clothes, his style has faded as much as the color of his shirt.

"You ready?" Sam asks.

"Yeah. Let's get it."

"See you later, Rob," Sam says as we walk outside.

The four of us jump in a car to drive over to the bar, tension clearly permeating the interior of this white Chevy sedan. Abby hands me her phone.

"Pick a song," she says.

I turn on a G-Eazy song. "Who's this?" Jamie asks.

"You've never heard of G-Eazy?" I ask incredulously.

"No. Is he popular?"

"Duh. Do you listen to any rap or pop?"

"Not much. Mostly country. That's all they play at this bar, just so you know."

"I see," I tell her.

We are in the car for a half hour, passing countryside and farmland until coming to a secluded bar in the middle of nowhere.

"IDs," says the bouncer before we enter the venue filled with blaring country music.

Once inside, I notice everyone is wearing cowboy boots, and most guys have their jeans tucked in and are wearing a dorky hat. This isn't my scene, but for that exact reason I will thrive. The numbers suggest that one woman here, no matter how much mudding or lassoing she does during the day, will be fond of my skinny jeans and formfitting shirt.

"I told you there would be dancing," Jamie yells over the noise.

"Why is everyone dancing together though?" I ask.

"It's called line dancing. You've never done that before?"

"I've heard of it, but most places I go to just play club music."

Jamie laughs. "This isn't that kind of place."

We move toward the bar.

"Can I get you something?" the cute but cranky bartender asks me.

"Grey Goose soda, please."

While sipping on my drink, I casually take glances at the various women in the bar. Sam is already taking the last swig of his second Captain Coke while fully engaged in a game of pool with Jamie, and Abby has migrated to the dance floor with a group of other friends.

"Yep, this is awkward," I mumble to myself.

I turn around and ask the bartender for a Coke, the desire to get wasted no longer appealing. Abby keeps throwing me glances as she dances in a circle with her friends, but I don't act on her invitation. With Sam in the good graces of Jamie, the future of my night is becoming clearer.

"How are we doing?" a joyful Sam asks me after concluding another game of pool.

"I'm having fun," I lie.

"Do you play pool?" Jamie then asks me.

"Unfortunately, no."

The mood of the bar does not match my preference. Going home and crawling into bed sounds like a dream. I could pursue Abby, but that would require an effort I don't want to give. But I also don't want to give up on the night.

"Are you sure you don't play pool?" Jamie asks with a friendly smile.

"Fine, I'll play a game of pool with you," I tell Jamie, standing up from the barstool and grabbing a cue stick from the corner of the room.

It doesn't take long for Jamie to easily beat me, but now I am a part of things, even if it's only superficially and part of a manipulative ploy to enrich the outcome of my evening.

Sam, Abby, Jamie, and a group of people with faces I'll never see again move outside to sit on a wooden bench and smoke cigarettes.

"You don't dance?" one of the strangers asks me.

"I actually love to dance. I just don't know the dances or music they're playing," I state. "This doesn't seem like a place you can just go out and dance."

"No. You have to know what you're doing," her friend admits, slowly moving away from me.

"And the music is pretty lame," I punctuate the end of our brief interaction with, but by the time my words are spoken she has already leaned into her friend for comfort.

I see Jamie standing to the side and want to change my fortunes. "Can I have a word?" I ask her, and then we move away from the group. Sam has noticed and looks on curiously.

"What's up?" Jamie asks in between puffs of her cigarette that she blows out the side of her mouth.

"I'm just curious. Are you guys into us? Like, are you attracted to us?" I ask.

"What do you mean?"

"Well, is Abby into me? Are you into Sam? Like, what's going on here?"

Jamie inhales another whiff of cancer and then smiles. "I find you attractive," she says.

"Hmm. What about Sam? Aren't you into him?" I needle.

"He's cute, but I definitely think you're hot."

Sam is now in jeopardy of not getting laid tonight. Morality is beckoning me to make a choice. I don't want to screw over my best friend, but a realist would tell me his luck has already run out.

Sandwiched between the bro code and my own wants, I defer making a decision. Instead, I segue.

"What about Abby? Is she into me?" I ask.

"She thinks you're cute, but she is kind of seeing someone."

"Kind of? What does kind of mean?"

"They're not together. I know she likes him though."

Frustration creeps in. "Jamie, I don't mess around with women in relationships."

"I wouldn't say she's taken though."

"Well then you're not really saying anything at all, and that's fucked-up."

Momentum is draining from the conversation. I peer over Jamie's shoulder at Sam, who is still rightfully interested in our void from the group. My gaze reverts back to Jamie and I go to a place where only fantasies live.

"What if we had a threesome?" I ask Jamie.

She chokes on the nicotine.

"You, me, and Sam," I say.

Jamie's eyes intensify.

"We can take turns, or we can double-team you," I say with the utmost sincerity.

Jamie takes another puff of her cigarette while contemplating the idea, then responds. "I'd be into that," she reveals.

"Then it's settled. I need to take a piss. I'll catch you later."

Sam catches me as I walk to the bathroom. "Where you going, dude?" he asks.

"Follow me," I say, grabbing his shoulder.

In the bathroom, I begin urinating and talking simultaneously. "Dude, so I've just set us up to have a three-way with Jamie," I say, immediately realizing I should not have led with that.

"What!" Sam cries, taken aback, his desire for a two-person romp with Jamie exiting the vicinity along with the sparse amount of alcohol in my bloodstream.

"It hasn't been decided if we are going to take turns or double-team her, but she wants to have sex with us both," I continue.

"Jesus Christ, kid," Sam groans in his here-we-go-again voice.

"It'll be fun, bro."

"Fun? I...I got to get back out there," he stammers, his voice trailing away along with his body.

I take his words to mean he is on board.

The mood has suddenly shifted. I leave the bathroom and head to the dance floor to jump into the next song like line dancing is what I was born to do. Even though my hips are completely out of tune, I still manage to make it through three minutes of guitar-themed beats and repetitive lyrics.

But when the next song comes on, my good fortune eviscerates. Flailing elbows and gyrating hips push me around the dance floor. All the cowboys and girls in jean-platted skirts look sideways at me. It's a reminder that this isn't, and never was my scene.

To walk away would mean failure, so halfway through the next song my chest catches a people's elbow from a broad with sweat oozing from her temple. On the next turn, she grabs my waist and tries to direct me where to go, but I'm more lost than a priest in a strip club.

I don't even let the song finish, instead putting my head down and shamefully sauntering back to the same barstool where I spent the early part of the evening.

Over time the bar clears out and my patience is also vanishing. Sam, Abby, and Jamie say goodbye to their friends and then we pile back into the sedan.

"So what's it going to be, Jamie?" I demand as soon as the four of us whip out of the parking lot.

Jamie and Abby trade eye contact, unsure how to deliver their next words.

"I don't know, Q. Abby has to get up early in the morning," Jamie defers.

I bristle. "Okay, but we already talked about this, Jamie."

"I'll be going home tonight, *alone*," Abby interjects.

"What about you, Jamie?" I desperately ask.

"Hmm, I don't know."

"Well, what about that talk we had at the bar?"

My tone is becoming more aggressive and unattractive. A few seconds pass and no response. "Jamie. What's changed?" I ask.

"I don't know," she finally uncorks.

The rest of the car ride home is silent. I don't dare look at Sam, fearful he doesn't know the whole truth and will think I've hijacked a good time from him.

Jamie drops Abby off at her home, and the three of us go back to Rob's place. I go to the bathroom on the main level and return to find Sam and Jamie standing in the kitchen with beers in their hands. A new proposition has emerged.

"Do you want to go to the neighbor's house and keep drinking?" Jamie asks.

"I'm fine. You two go ahead," I say.

"Q, are you angry about something?"

"Not really. I just don't understand what's changed."

"Let's talk about it over there," she says.

"I'm okay," I say, cementing my feet in the carpet.

Sam really wants to keep partying. "He's tired. You and I can go," he says.

Jamie looks at me with concern.

"Seriously, I'm not angry," I assure her. "Just disappointed."

She frowns. The two of them then leave, and I crawl into the daughter's room where a pink unicorn blanket traps all my body heat. Plausible deniability has robbed me of pleasure.

CHAPTER 12

In search of an itinerary for the day, I walk out to Rob's school bus the next morning. Sam is inside, scrunched together in a deep slumber inside a sleeping bag.

"Bro, what's the deal? We got to get moving if we are going to ride today," I tell him.

Sam rolls over, not yet cognizant but still realizing the basis for my presence. "Rob said we can stay another day," I continue, although he was likely unaware of the previous night's events when he offered.

I'm not enthralled about staying another day in the same location, but I really like Rob. The dude loves basketball, has a pleasant face, and treats his family the way a man should. Rob exudes a greatness I can't currently match, but inside I want what he has. What I see in him, I hope I will one day see in myself.

"I'm going to have to call hosts for the next few nights and see if I can push us back a day," I inform Sam as he slowly crawls out from his sleeping bag.

"You're cool if we stay another day?" he mumbles through the morning chill.

"We aren't on a schedule, bro."

"I know, but you're a Nazi about schedules and this kind of stuff."

"We're good, bro," I smile. "Let's take the day off."

There are no outside forces constraining this trip, nothing that is forcing us to expedite our arrival in Maine. As long as the money situation is good, we can stay on the road for a long time, and so far, we have been living off less than ten dollars a day. A lot of that has to do with luck because many Warmshowers' hosts have provided dinner and sometimes breakfast, but we also haven't spent a lot of time in large cities where the cost of living is higher.

I exit the bus and go outside. I need to call our hosts and see if they're okay with us pushing back our arrival by a day.

"That's no problem," both people for the next two nights assure me.

"Great. I'll see you guys later then," I tell them.

With the day to ourselves, now is a good time to find a bike shop and get a prognosis on my chain. I haven't shifted gears since that moribund road outside Milwaukee, only able to get away with this because we haven't encountered steep inclines, but that's likely soon to change.

"Bro, we're good for the next few days," I tell a still-groggy Sam. "I got to get going though. I need to head over to the bike shop and get this chain thing figured out."

"I'll just be here, relaxing," Sam replies.

I grab my bike and ride a few blocks to the store. "Do you think you would have time to look at my chain? It keeps popping out of place," I ask the shop owner.

The man bends down and begins examining my bike. Within minutes, he has seemingly diagnosed and fixed the problem. He starts mentioning terms that are unfamiliar to me, and I politely nod my head because the verbiage he's using brings no stimulation to my brain. My only concern is that this annoyance with the chain becomes a thing of the past.

"That's incredible. How much do I owe you?" I ask him.

"Nothing," he says. "Just pay it forward."

Later that day Sam and I walk down to the beach. There is a lot of construction, so we zigzag through the sand until there is a clearing near a lighthouse where we can set down a few towels. Sam then begins undressing, revealing his enormously hairy physique.

"You coming in?" he asks.

"It's freezing, bro. I can't do it."

"Suit yourself," he says, and with that he bolts toward the water and dives in headfirst like a dolphin.

THE LONG ROAD EAST

Sam begins flopping around in the water like he's eight, giving me a chance to lay my head down in the sand, take a nap, and let the warm sun smack against my face. A half hour later, drops of water splatter onto my face while Sam dries off, interrupting dreams I'll soon forget.

"It's Kathy's birthday today," he says as he wipes off his wet armpits.

Sam always calls Kathy on her birthday. They are former lovers, and for years this has been an annual tradition.

"Why not just break the streak this year? I hate to be a dick, but she probably won't even notice," I tell him.

"I can't, man."

"I know, but it's got to end sometime. Why not now?"

Sam thinks about it for a little bit but then decides to go call her anyway. "Hey, Kathy! Happy birth...," I hear before his voice trails off.

Part of me knows Sam has some inner workings in place, a long-conceived plan he's hoping to execute. He is smart enough to see the annual exchanges with Kathy as strictly platonic, but the subconscious always keeps a sliver of hope that never vanishes.

I hold onto it myself, constantly thinking the woman I still love will one day call and forgive my infidelity so we can continue writing our fairytale narrative, minus a few key chapters of course. These thoughts may be futile, but they make each day a little more doable.

Later that night we join Rob's family and go to one of his softball games. Much to his displeasure, Rob strikes out every time he comes to the plate.

"I usually do much better than that. You guys are a curse," he jokes.

Afterward we drive to a house for what is supposed to be a family-friendly gathering, but Rob's wife and kids soon leave, and now there remains just a handful of guys. Sticking around without the presence of women isn't enticing, but as the sun sets, beer and cigars are being handed out like candy. Sam's gregarious personality immediately becomes a hit with all the guys, but when our previous night's events are brought up, describing them is handled with less than graceful transparency.

"I got this kid over here"—Sam says, pointing my direction as he holds his second can of beer—"telling me we are going to double-team Jamie. And I'm just like 'what the fuck, man?'"

The audience is intrigued. "I mean, here's Q," Sam continues. "He's trying to fuck anything that moves. And here's me, I'm a four, and I'm just trying to figure out where I fall into all this." Sam laughs.

"Don't worry, Sam. I'm a four, too," interjects a weird guy named Ruxin who gives me the creeps from the jump.

Ruxin has this devious smile, and his body shakes back and forth like he has Tourette's and a bad heroin addiction. He also holds his smiles too long, the reception changing from nice to overbearing in a matter of seconds.

"Yeah, yeah," I answer each time Ruxin says something, hoping that if I'm dismissive he won't ask me any more questions.

Soon out of beer, one of the guys runs back to his place to grab more. His definition of more includes bringing back a small keg. It comes with a nozzle, reminiscent of the setup the Javier Bardem character has in *No Country for Old Men*, the one he totes around and uses to shoot everyone.

Whatever kind of beer is in this keg takes the entire evening to the next level. Soon Sam is puffing on a cigar, Rob is crying from laughter, and Ruxin has broken a chair. Needless to say, people are getting *fucked-up*.

"I was trying to buy weed one time in Mexico, and all I had was a couple bucks," Sam starts, preparing to tell a story that will forever cement him in Grand Haven's suburban lore.

"It was nighttime, and I'm walking along the beach, you know, just trying to find some weed. I get to this dock next to the ocean, and suddenly I hear this voice. This voice is calling to me, but I can't see because it's so dark out."

"*Hey, hey. You want to buy some drugs?*" the man in the story whispers.

"So I look around, like is anyone watching me, and where did this dude come from? I get closer. And, fellas, I am not shitting you, it was so dark I could only see the whites of his eyes."

"*How much money you got?*" the man in the story asks.

"A couple bucks."

"A couple bucks? He probably gave him a bag of shit!" a dad blurts out, causing another roar of laughter from the gallery.

"*Alright, give me your money.*"

"I paid him, got some weed, and smoked it on the beach. Didn't die or get arrested. Pretty good night," Sam finishes with a smile while the dads rock sideways in their chairs trying to contain their robust laughter.

They can't get over the fact that Sam bought weed from a random stranger underneath a Mexican bridge, and I can't understand how impeccable Sam's ability to make random strangers admire him is.

As the dads recount the story and transition into talking about coffee beans made from the defecation of bats, Sam wears a proud grin on his face, a deserved one because no matter how much I try, I will never have a personality that can adapt to every circumstance.

Everyone begins exiting when the pulse of the midweek party begins to simmer. We need a ride back, and the only willing provider is none other than our new pal Ruxin.

At this point, Ruxin has already collapsed and broken a chair. He's clearly inebriated, but no one is concerned with anything other than getting back to the house. We all pile into Ruxin's green car, still slapping our knees from all the laughs elicited by Sam's convivial personality. Rob and I force Sam to sit in the front.

"Be careful. He might just ask you for a little hanky-panky," Rob quietly chirps at Sam while Ruxin limps around the trunk to the driver's side.

"We all ready?" Ruxin asks, turning the ignition over once everyone has strapped on their seatbelts.

Before we begin moving forward, Ruxin slams into the curb.

"Are you sure you should be driving?" Rob asks, now switching to dad mode.

"I am *fucked-up*," Ruxin responds with a weird smile. "But I can drive."

Then we pull up to a stoplight. "You know, I haven't gotten laid in five years," Ruxin shares.

"Neither has Sam!" I shout.

"Really?" Ruxin asks Sam with delightful surprise.

"Don't listen to this kid. He's an idiot," Sam booms, the strands of hair on his arm sizzling with rage.

"You know, Sam. Us fours, we have to stick together," Ruxin says in a show of solidarity.

Rob and I burst into more laughter. One more Kumbaya and I'm sure Ruxin will whip out his dick. We then stop off at a gas station. Sam picks up some beers for a nightcap, and Ruxin goes in to buy cigarettes and cool off.

"Sam, I'd be careful, man. Any more, and Ruxin might just follow you into the bus tonight," Rob cackles.

"That guy is fucking weird," Sam complains.

When Ruxin drops us off ten minutes later, the three of us barrel inside as fast as possible. I hear Ruxin bottom out when he backs up, a likely sign he didn't squirm into the bus.

CHAPTER 13

"I think I'm still kind of drunk," Sam says the following morning as he cracks open a warm Blue Moon that hours prior didn't make its way into the fridge.

"It was a real pleasure having you guys," a groggy Rob groans with as much enthusiasm as he can muster when he comes down the stairs.

"It was a pleasure staying here," I say.

Jamie walks upstairs from the basement and joins us in the kitchen. The weather outside looks anything but inviting, but it's time for us to leave.

"Where were you last night?" Sam asks Jamie.

"Some things came up," she lamely answers.

"It's too bad we couldn't make something happen," I say in a lower decibel.

"What do you mean?" Jamie deflects.

I don't even bother responding.

"Anyway, it was really nice to meet you guys. I hope you stay safe," she continues.

"Tell me something, Jamie," Sam says as he rests his half-empty bottle of Blue Moon on the counter. "Are women here all talk too? I thought that was just a Minnesota thing."

Sam's forwardness throws Jamie.

"Uh, I have to go get ready for work, and, well, um, I feel like shit," she says, rubbing her forehead and then walking back downstairs and out of our lives.

Sam and I are leaving Grand Haven full of memories and laughs, but not without a lesson.

"Bro, in the future, we can't get attached like that," I tell Sam.

"What do you mean?"

"What I mean is that we are going to be entering and exiting homes very quickly on this trip, and if we develop feelings, it is only going to lead to pain."

"I know, but that family was so cool."

"They were, but there will be a lot of cool people on this trip."

I snap the straps on my backpack together, and we ride off. Quickly Sam pulls ahead as we jump on a trail to get outside of town. I am on top of a hill when suddenly my chain jams again.

"What the hell!" I scream, looking down and seeing that the chain is barely getting through each cycle.

I desperately revert back to the gear I was in, and the chain fortunately begins making full rotations. My mind goes right to the guy at the shop, irritated he didn't properly fix my issue. He did help for free, but in many ways, he didn't help at all.

Rain is now coming down while riding past middle-class homes. I'm about to cross the street and keep with the trail, but then a red minivan darts through its left turn and slams on the brakes when I'm in the middle of the road. The rotund woman inside blares her horn and flashes her middle finger.

"Fuck you, you fat bitch!" I yell, incensed she can't wait two seconds for me to pass.

I continue on to the other side of the street. The woman speeds past as if slamming on the accelerator will forgive her attempted trespassing of the crosswalk.

"You won't believe what just happened," I tell Sam when I come up on him in the middle of a wooden bridge.

"What happened?" he asks, but my story isn't his primary concern. He is monkeying with one of the bags on his bike.

"Is everything okay?" I ask.

"It's fine. I fell."

"Did you slip because of the rain?"

This question brings no answer.

"Fuck. This thing is not going to get back on," Sam laments, wiping the rain from his brow.

THE LONG ROAD EAST

My anxiety heightens, fearful we've reached a point in the trip where pesky mechanical issues become the norm. We have never ridden this many miles. How the bikes will hold up over the long-term is a mystery.

"This will do for now," Sam says a couple minutes later once he jerry-rigs the bag onto the frame of the bike.

When we stop at a gas station, I buy a couple donuts and some chocolate milk. Sugar and I are becoming fast friends these days, a newfound fetish I justify by the amounts of calories I'm burning. Sam is unimpressed.

"Those donuts good?" he asks in the way people do when they wish their metabolism matched their intake of processed meals.

"Sort of," I admit, not trying to let on that the treats taste glorious.

If these donuts ever become a gateway drug to candy, I'm screwed because in high school I used to consume Skittles like there was only one rainbow left in the entire sky.

"Let's see what you buy then, bro," I tell Sam.

He comes back out with a pint of Fireball, his way of masking the repercussions of last night's good time.

"You want a sip?" he asks, taking a swig and smacking his parched lips while extending the bottle.

"Nah. I'm good."

We keep riding, our day slowed by relentless rain that won't go away. It's been a shitty ride. Both our bikes have taken a beating, the route hasn't been enjoyable, and as we position ourselves at a left turn, another annoyance enters.

"Hello, excuse me," comes a woman's voice from a car two lanes to the right. I crane my head to see a middle-aged bird with a frumpy mug sticking her beak out the driver's side window of an SUV.

"My husband is a biker, and there are hand signals to use when you're crossing lanes of traffic," she says.

"Good for him," I want to yell at the old hag.

I am aware of the hand signals to which she is referring, but they're the last thing on my mind. I'm more concerned with getting some food and laying my head down on a soft pillow, but what I think doesn't matter because a traffic violation has occurred in front of an importunate individual.

"Are you having a good day, miss!" Sam yells, a few drops of saliva exploding from his mouth.

The woman jiggles in her seat. "Look, I'm just trying to help," she claims.

"I said, are you having a good day, miss!" Sam demands to know once more.

She scoffs. "Geez. I only asked a question."

An uncomfortable silence ensues. A green light would end this painful moment. The woman in the car looks like she is about to scream, but nothing comes out. She begins pouting while everyone in the vicinity awkwardly waits for the light to turn green.

She keeps looking back and forth at Sam for an apology that isn't going to come. The light doesn't change for another thirty seconds, and I can't help but laugh. It's been one of those days.

"I don't think she was trying to be rude," I tell Sam as we get back on a trail and close in on our destination.

"I know, but it was so unnecessary to say anything."

"I agree. I think she might have had some other problems going on, and that's why she was so concerned, ya know?"

"Q, I don't give a damn."

In truth, I think the older woman just wanted to be heard. Maybe because her husband didn't give her the attention she wanted, or maybe she was an intrusive person. Either way, I don't know how that woman expected us to respond. She couldn't possibly have thought we were going to jump with joy at her comment, eternally grateful for information we could use for the rest of our lives. The truth is you just never know where people are coming from.

After a series of wrong turns, we pull up to June's house around 9:00 PM. Most of the lights are off, and I am worried she's given up on us. I ring the doorbell twice, a light then comes on and a figure begins walking toward me.

"Hi, June," I whisper as she slowly opens the door. "I'm Quentin Super. This is Sam."

An elderly woman, closer to the end than the beginning, June lets us inside and then starts pulling leftovers from the fridge. As she warms the food up in the microwave, she begins sharing tales of her biking adventures, a list that does not impress the hypercompetitive Sam.

"She's kind of a poser," he says, mentioning the fact that June rode in large groups to avoid the wind rather than bury her head in misery like us sad cases.

"She's an old woman," I laugh. "What do you expect?"

I know Sam is stubborn, but this is next level, trying to best a woman who looks like she might croak any minute. But then again, you never know where people are coming from.

CHAPTER 14

I don't know what it is about people who work in bike shops, but 97.6 percent of them are total assholes. Bike shops are the only stores where nearly every employee is condescending to the customers.

You don't go to a hair salon and get berated for not knowing how to cut hair. You don't go to a lawyer and get scolded for not knowing the subtle nuances of mortgage laws. People in bike shops are the only people on this planet who talk down to the people trying to pay them to help fix a problem.

"You are on a tour, and you don't know the air pressure of your tires every…single…day?" a balding man with glasses chastises after I politely request that he put air in the tires of my bike.

"No," I state, elongating my response with emphasis, one more snide comment away from informing this loser with a beer gut who works for eleven dollars an hour that he is egregiously rude and that I'm not hanging out with him by choice.

"What's the name of the shop you took it to in Grand Haven?" a more reasonable employee comes over to ask.

"I don't remember," I say.

"There are two."

He lists off two names that don't sound familiar.

"I honestly couldn't tell you," I reassure.

"I only ask because they did a poor job the first time."

"Damn. Well, he did do it for free, and I was out of the store in five minutes," I say.

"But you don't remember the name?" the guy annoyingly asks again.

THE LONG ROAD EAST

"Like I said, he did it for free. He was a nice guy, and I'm not really in the business of throwing people under the bus."

Then, a mechanic from behind the counter chimes in. "Yeah, whoever worked on your bike did a real *shoddy job*."

"That's great," I mutter.

All I want is to get my bike permanently fixed so I can leave the store and never come back. To temporarily escape the confines of these suffocating pricks, I walk over to Verizon to see if they can replace the battery in my iPhone.

"We don't do that here," a woman says as I stand in the lobby with an armful of bananas and a donut that might have been fresh a few days ago.

"So there's nothing you can do?" I ask.

"That's correct, sir."

I walk back to the bike shop and inhale the glazed donut.

"Is my bike almost ready?" I ask one of the workers.

"Definitely, and I'm only going to charge you sixteen dollars," the more reasonable man from earlier tells me.

With my bike problem behind us, Sam and I begin the day backtracking two miles. Over time, the rain worsens into a torrential downpour. Stopped underneath an awning in a small town, we admire the sheer velocity of the rain that's crashing down against the road. My clothes are soaked, and for the first time all trip I'm concerned about catching a cold.

"We're on a trail for the next fifteen miles," Sam authoritatively states.

"After you, bro," I say, and then we both plunge forward into the scissoring rain.

I'll follow Sam to the death so long as I get to be present when he unleashes another diatribe on an unsuspecting Karen.

We zoom through the rain even though it slaughters us, pellets crunching against my jacket and droplets buzzing into the nape of my neck. As we move along the trail with speed, the thunder begins hijacking Sam's momentum just enough so that we are able to ride parallel for a few miles. The trees on each side of the trail are slightly protecting us from the interference, and it seems like we might just get through the day in decent shape. Then the pavement of our trail suddenly ends.

"The GPS says to keep going," a confused Sam alerts.

"That trail up ahead is grass though."

"I don't know. We might as well try it. The road is too dangerous right now."

The figure of a man riding a horse is plastered on a brown sign just ahead of the trail. My tires sink into the wet grass when I first enter, relegating me to a slow crawl. Predictably, Sam fares better and speeds ahead. I trudge forward, cautiously avoiding tree branches, dog shit, rocks, and holes in the ground, doing everything I can to avoid getting a flat or popping my chain out of alignment. When I catch up with Sam, he is parked in front of a downed tree.

"Jesus fucking Christ!" I wail, the sheer sight of the downed oak enough to send me into a tailspin.

"There is a small opening over there," Sam says, pointing to the left at a sliver of hope that has presented itself in the form of a few dainty tree branches.

We take turns wedging our bikes through the maze of branches and wet dirt. I lift my bike off the ground to avoid getting mangled in a pile of goop. My soft green shoes are losing their shine to all the mud sucking at my soles. A sharp edge on one branch catches me in the face so I rip it off in frustration.

Joad is lumbering behind and my fingers start to sour. Her wheels are falling into every crevice, and part of me wishes nature would take her off my hands, suck her up like the lava from one of those weird *Spy Kids* movies. I want a break from my dependent, the reality of dragging Joad across three state lines starting to wear.

On the other side, the tree is still situated in the same sedentary position. "I probably have bugs all over me now," I complain.

"We'll be out of this soon," Sam assures.

We're about to reach the road, but just then the pesky chain on my bike falters. I slam my front tire against the grass, irate.

"And that asshole had the nerve to say the other guy did a *shoddy job*," I disgustedly yell at no one in particular while Sam pops my chain back into alignment. "I'm going to call them tomorrow and bitch those fuckers out."

"It is kind of fucked. Like, everything that's happened," Sam admits.

We are now back on an actual road, but there is no shoulder. Cars either can't see us or don't care about safety because they zoom past and leave only a few feet between human skin and their spinning wheels. At a right turn the GPS wants us to go down a gravel road.

"You have to be fucking kidding me," I cry, thinking Google Maps has been hacked by the Chinese.

A few moments after the gravel road we have to cross a busy county highway. Our vision is blurred by the pounding rain and gradual mist that permeates the chilled air. Cars have their lights on, but the precipitation renders them virtually useless. In one of the cars sits a young boy in the passenger seat. His face says it all: *thank god I'm not in your position.*

So much for Sam and I inspiring people on this trip.

I sneak closer to the road, trying desperately to get a sight line to the oncoming traffic from the right. Hailing rain bounces off the road and distracts from the fact that I cannot see more than fifty feet ahead. It's a crapshoot as to whether cars are coming. The cars next to us remain stationary, so I wait for Sam's guidance and look up. The dark skies are intent on not showing mercy anytime soon.

"This is so fucked," I huff through gobs of rain that fall down my face and into my mouth.

The danger is clear. There is no way cars can see us while crossing this road. Even if they could, going downhill at sixty miles per hour during a downpour is not an ideal time to slam on the brakes.

"We are going to have to go with one of these cars," Sam says, and it's so cold I can see his breath. "Wait until one of them goes."

A few more moments pass. I suck more water from my CamelBak and grip the handlebars with more force than the woman who gave me my first hand job.

The few cars to already go through the intersection had to dart through. An ideal opening is not going to come. Sam and I have to make one. The next car in the queue lurches out, and we are going with.

"Go, go, go," Sam neither yells nor whispers.

Our safety is no longer in our hands. I quickly swivel my head both directions, lights from opposing cars visible but their distance unquantifiable because the only thing my mind is focused on is getting to the other side. My muscles are clenched, braced to get pulverized from either direction.

But it never happens.

Later that day, a few miles from nondescript Nashville, Michigan, the rain has cleared, and the sun is beginning to show. We take one final break next to a church. A group of prom-bound teenagers are funneling inside.

"Would you ever have sex with a chick whose still in high school?" I ask Sam.

"Hell no! Would you?"

"I mean, if it was legal, for sure. That would be superhot."

"Q, that's how you go to jail."

"Dude, I'm not saying I'd do it. I'm just saying it'd probably be fun."

Sam looks at me with consternation.

"Bro, chill out. Remember, I never got laid in high school," I tell him.

"Don't tell anyone you said that," Sam advises.

"Which part?"

"Both," Sam nods before speeding ahead, and we soon arrive at the Warmshowers.

"Come inside. Come inside," our host Uma says, ushering us into her home. "Leave your wet shoes outside. Art will be home soon."

Uma's home is a small cabin with wood interiors, granite counters, and a fireplace sectioning off two bedrooms. A few minutes later Art comes through the front door with a big smile on his face.

"Eat anything in the house. Don't ask, just eat it. I'm serious," he says after getting settled.

I throw some dirty clothes in the washer and sit down on the couch to relax.

"You guys want a beer?" Art asks as he pulls a few Coronas from the fridge.

"I'll have a beer with you," Sam says in his customary way.

"Perfect. I only have three every year. This seems like a good time for the first," says Art.

Our hosts have prepared a fantastic spread, full of meat, protein, and fruit. I try to make Art happy and eat everything, eventually going so far as to dig out some Rice Krispies bars from a drawer just below a vase of fruit.

After dinner we sit down in the living room to relax. Art begins to share memories from his past.

"I hadn't paid my taxes in twenty years," he leads off one of his stories with. "I had this wealthy uncle, and when I told him I never pay taxes, he flipped, telling me I had to get a lawyer and that they might throw me in jail. I was only making thirty to forty thousand dollars per year, so I didn't think anything of it at the time. Then one day I get a letter from the IRS. I go down to one of their buildings. They lead me through all these doors. I mean, they're treating me like I'm a bank robber. I'm scared shitless because I'm thinking I'm going to be going to jail for a long time."

Uma watches Art tell his story. Even though she's heard this tale a thousand times, it still excites her as much as if it were her first time hearing it.

"I'm sweating balls, thinking my life is over," Art continues. "Finally, a guy comes in and tells me I'm free to go. I'm thinking no way. They're not going to bring me down here, put me through all this and tell me I can go."

"What happened?" I ask.

"I asked them if I could get this in writing because I was going to go straight to the bank and get a loan. They put it in writing, and I left. I called my uncle after to tell him the news."

"*Pay your damn taxes, Artie!*" the uncle demanded.

This sends Sam into a tailspin of laughter.

"But what about you guys? What's your story, Q?" asks Art.

"Well, Art, I'm a broke college grad. Just got out of a relationship. I guess in many ways I'm miserable," I joke.

"You're young. You'll get through it."

"Do you think I should get married?" I ask. "I feel like all I do is chase women around. To be honest, it's kind of exhausting."

"Oh, you're going to be following that *thing* you got there until you're at least forty years old."

"That can't be good," I mumble.

"Look, we have to go to church in the morning, but seriously take whatever you want when you leave. Don't worry about money or anything like that," Art says before slipping off to bed.

As I curl into bed later that evening, the light from my bike inexplicably starts flashing outside the window.

"I'm not going to be able to sleep with that thing going off," Sam hints.

"For real?"

"Yeah, man. For real."

"It's raining though."

"Q, get up and go turn the damn light off!"

I walk into the pouring rain. The button on the light won't click off, and each passing second exposes me more to the unrelenting storm.

"Holy fuck," I say, shoving the light in my bag on the back of my bike. "Will this day just end already?"

CHAPTER 15

Mist and fog blanket the landscape, setting an ominous tone for the day. I eat too much for breakfast because right now I am going through so many calories that never get replenished. The science behind my metabolism eludes me. I only know that if I don't eat every hour, my stomach will start barking more than a dog left in the rain.

We're only a few miles in when my bike takes a turn for the worse. The chain's popped out of alignment again, and this time I can't gear back into place. Sam is already too far ahead, so I frantically whip out my phone.

No service.

Just like that, the first potentially damaging position of the trip. I fiddle with the chain for a few minutes, trying to get it back into alignment. I keep yanking, but my scrawny arms are no match. I start walking, hoping to find signal before Sam gets twenty miles in.

It's easy to fall into a panic. No phone service, the incessant rain, and the stupid bike mechanics in Kentwood that talked a big game and wound up being utterly inept. All these things make me want to rip off the shop owner's glasses and smash them to pieces.

While walking and mumbling to myself, I pass a tarnished white house with a creepy middle-aged woman covered in dirty clothes. Fields surround her property that has no neighbors. She's strange to look at, so I start scanning the encyclopedia of movies stored in my brain for horror films where the city kid gets skinned alive by the cannibal farmer.

She yells. "Hey, do you need some help?"

I stop and give her a quick turn of the head. A few more grungy specimens have taken a seat on her front porch. I don't stop walking because they look hungry.

The stereotype is a homeless guy will pull a gun on you in a dark alley late at night, but in my opinion rednecks are the scariest subset of humans on this planet. If you're in a city or suburb someone might try to take your money in white-collar fashion, nothing too intrusive. But if you get out in the sticks, watch your back. It's easy to imagine a deranged farmer coming up from behind with his pickup and knocking me off my bike, then bringing me to his lair and feeding me to the pigs.

See, if someone mugs you in the city, there are people and cameras everywhere. Here in the middle of dilapidated Michigan, no one will see this crazy woman who needs a bath stab and drag me inside her home.

"No, I'm okay. I don't need help," I yell back.

Once they are out of sight, I start fiddling with the chain again. It still won't pop into place. Seventy-five miles to go, and I have no clue how far ahead Sam is.

"Fuck this, fuck this, fuck this," I keep saying in disgust.

My mind keeps harkening back to the idiots in the bike shop, and then I start thinking about the guy from Grand Haven. I wonder how my bike went through so many hands without coming to a resolution. I could care less about the money. It's that right now I'm vulnerable. No cell phone signal, and no stores or gas stations. Just a country road with a few intermittent farms and the threat of a wild boar pig lurking in the distance.

And then for two seconds I think that death would be a welcome sight. I hate myself for blowing my relationship. I hate being addicted to promiscuity and Tinder and chasing women to all corners of the earth. I just want to feel normal, happy to go home to one woman and not be forced to follow my dick around until I'm "at least forty years old."

Enraged, I get down on my knees again. I'm going to yank the chain back into place, even if it means scraping all the skin off my knees.

THE LONG ROAD EAST

Just then, a sedan pulls over on the other side of the road.

"Do you need help?" a woman from the driver's side asks as a man simultaneously gets out of the passenger seat.

"Do you have a phone?" I ask, sensing this is my last chance to take someone up on their offer for help. "I have no signal, so I can't call my friend."

"You can use mine," the man offers.

I dial Sam, and he fortunately answers.

"I'm on the side of the road. Some people let me use their phone," I tell Sam. I look around at my surroundings, and there aren't many options. "Fuck, dude. I think we might have to get towed to the next stop," I continue.

"We have to get going to church," the man says when he can sense the disharmony in my tone.

"Okay. Sam, I gotta go," I say, then begrudgingly hand over the phone to the man.

The car takes off, and I'm alone for the next twenty-five minutes until Sam thankfully returns.

"I could probably pop the chain back into place again," Sam says optimistically.

"This thing is busted, Sam. And I don't know what the fuck to do. No one can fix it. I'm going to go broke just trying to fix this one problem."

"You could at least try to make it a certain distance before we call a tow," he says.

"I don't know. The roads are getting hillier. I have to be able to switch into an easier gear, otherwise I'm going to get worn out too quickly."

Sam rolls his eyes. He will be upset if we throw in the towel, but realistically, my bike can't go on in this condition. The thought of popping the chain back into place all the way to Portland makes me want to gag.

I use Sam's phone to dial up my insurance company. Many companies will tow bikes, and depending on how far you have to go, they will do it for free as part of your policy. Oddly, I know this because we were in this same position not too long ago.

It was the spring of 2016. Sam, Chad, and I were taking off from St. Cloud, Minnesota, and headed for barren South Dakota. The first day was a blast. The sun was out, and we got to a small town just after lunch. We were only looking at another forty miles to our destination, another petite village by the name of Granite Falls. Getting there would mark a one-hundred-plus-mile day, a feat we took much pride in.

Customarily, Sam led the way that day. I followed behind, and then Chad took his sweet time in the back. This was how our system usually operated, and we didn't get too bothered if the distance between us lengthened throughout the course of the day because we all ended up in the same place, sitting at a bar telling raunchy jokes.

About twenty miles away from Granite Falls, I found Sam on the side of the road, his bike inoperable.

"What's going on, man?" I asked after seeing his new predicament.

"Some spokes blew out," he said.

"What should we do?"

"You guys go ahead. I'll have to think of something."

Chad and I reconvened up ahead to deliberate. "We could maybe book a hotel here," I said, looking around an unincorporated city and seeing only one quasi motel and a massive power plant. "There isn't much here though."

"We have to get to Granite Falls," Chad said.

"What about Sam? He can't walk his bike twenty miles."

Chad then called Sam. "Where are you? We are trying to think right now."

"I'm actually in the back of a highway patrol car. He's bringing me to Granite Falls," informed Sam.

Only a guy like Sam possessed the charisma to somehow get a lift from a cop *and* be chauffeured the rest of the way. So Chad and I busted our tails to meet him.

The next morning, the three of us sat in our hotel room contemplating how to solve Sam's dilemma. There were no bike shops in Granite Falls, and the next closest one was in Marshall, Minnesota, a college town thirty miles away.

I called the only taxi company in the area.

"How much would it be for a taxi from Granite Falls to Marshall?" I asked a man.

"Around one hundred dollars," he said.

"Damn, that's a lot. I'll call you back," I said but never did.

When I mentioned how much the taxi would be, the boys bristled. "What a fucking joke," Chad said.

"I can try AAA. They do bikes, I think," Sam speculated, then going to work on the phone. AAA initially claimed Sam didn't have bike coverage because the insurance was based out of South Dakota.

"But I'm only in Minnesota. This is a national service," Sam rightfully protested.

"Bikes are only covered in the primary state of the account holder. I'm sorry, sir. I cannot help you," a woman with an Asian accent told him.

Suddenly, Sam's impromptu declaration, from the comforts of a hotel, saved our ride.

"Look, I'm stranded on the side of the road! It's raining out, and I might get hit by a car! Do you want that on your hands?" he yelled at the poor woman.

"One moment, sir," the now-frightened woman said.

Chad and I did everything in our power not to burst out laughing, and Sam kept waving his hand for us to shut up. Soon my stomach began to hurt from all the repressed laughter.

"Sir, are you okay? Are you there?" the woman on the other end asked with concern when she returned.

"Yes, I'm here," Sam said, feigning agony.

"I've had a long talk with my supervisor. She says it's my call, so I'm going to send out a tow because I want you to be safe."

"The tow has to fit three people and three bikes," Sam just then informed her.

"Three people?"

"Yeah, I forgot to mention that me and two other people are stranded," Sam winced.

"You didn't say this earlier," the woman said with suspicion.

"I forgot."

There was another pause as the woman went back to her supervisor.

"Don't screw the pooch, Sammy," I warned him.

"Hello? Sir?" the woman came back with.

"Yes?"

"Is everyone there a AAA member?"

"Two of us are."

"Geez, okay. My supervisor says to do what I think is right, so I'm going to make sure you guys get picked up."

"Thank you so much. We really appreciate it," he said, ending the conversation, after which we all burst into a boys-will-be-boys type of laughter.

I thought the madness was over, until our ride showed up. It was an ambulance from the 1960s that looked like it had been pulled from a junkyard. The three of us shoved our bikes inside, knees crunched against metal, and the two guys in front not entirely sure what they were dealing with.

I put a photo on Facebook that suggested one of us was in this ambulance because of a horrific injury. And then the concerned citizens of Facebook began flooding me with messages. Some things you just can't script.

Back in Michigan, karma is catching up.

"We can only cover up to forty miles," the woman from the insurance company tells me over the phone. "You have to pay the rest."

"Even for bikes?" I ask.

"Yes, sir. Even for bikes."

"I suppose that will have to work then," I acknowledge, handing Sam his phone back.

The days of spending only ten dollars now seem like a distant memory.

"We look like homeless scrubs," I tell Sam.

I begin munching on a Pop-Tart. It doesn't ease the embarrassment of sitting roadside in the demoralizing rain that's made an appearance today and every other day we've been in Michigan.

A tow truck is supposed to come meet us, but forty-five minutes later he's still missing.

"Where is this guy?" Sam remarks.

Just then Sam's phone starts buzzing.

"What's your address?" asks the cranky tow truck driver.

"I thought my insurance company gave you the address," I say.

"They did."

"Okay, so what's the problem then? Do you have a GPS?" I ask.

"I do, but it's not bringing me to your location."

"That's weird, sir," I say, unsure of what exactly to say. "We are on the side of the road in the middle of nowhere. I don't think it'll be too hard for you to find us."

"Fuck, fine. I'm just going to have to call the insurance company again," the man whines.

Eventually, he finally finds us.

"Where are you going today?" he asks as he ties our bikes onto the back of his truck.

"Jackson, sir."

We step inside the backseat of the vehicle. A woman we soon come to find out is the man's significant other is sitting in the front seat. She is younger, cleaner, and more innocent than her partner. It appears she could do better, but based on the way she lets this man insult her, she must have her own issues.

"You're a real keeper," the man rudely and sarcastically tells her after she fumbles a pen from the glove box.

"You guys married?" I ask.

"Married?" the man laughs. "Fuck no. I'm not going to give her half my money when she decides to leave."

The man begins laughing uncontrollably, ignoring all sense of common human decency. He is a bigger asshole than me, and he keeps wiping his nose on his forearm, and then suddenly he wants to talk about the city of "Day-troit" and further showcase his lovely personality.

"There's fucking niggers running 'round everywhere up there. Go down the wrong street and 'dem niggers might rob you," he claims.

I look over at Sam, wondering if we are listening to the same guy.

"I used to love Day-troit, and then 'dem niggers ran everyone out," the man continues.

A few more of his vile sentiments come out, and I figure it's best to reserve giving a response because this man sees the world differently than I do.

"Jackson's full of 'dem niggers too. Some really good-looking ones, but they be getting pregnant when they're sixteen and whatnot. But damn they're some pretty niggers."

This guy has no sense of the room.

"And cops, man. They're crooked. They used to pull me over all the time and give me tickets on my truck. I never paid 'dem fucking things, so one day they dragged me into court, and I paid 'em."

"Why didn't you just pay the tickets right away?" I ask.

"That's how those cops is, man. You can get out of any crime if you got the cash. I had a buddy who got busted with a bunch of coke. Was looking at fifteen to twenty years. Told the fucker, now you're going to have to open up your wallet and pay 'dem fuckers."

"What happened? The court just let him pay and go free?"

"Yeah! I'm telling you. Ain't nothing in this world money can't buy."

We stop at a gas station to refuel.

"Can you bring us to Onsted?" I ask the man about a town roughly twenty miles ahead.

"For another forty dollars, sure."

"Why do you want to go there?" Sam asks.

"I've lined us up a new Warmshowers. If we stay here in Jackson, we'll have to pay for a hotel."

"But what about your bike, Q?"

"If you pop it into place I can get into Ohio riding with one gear. I'll bet they have a shop there."

"It's your call," Sam says.

I turn back to the degenerate man. "Forty dollars to go to Onsted, you said?" I ask.

"Yeah. I can do it for forty."

"Let's do it then."

We arrive at the house, and I am *praying* the Warmshowers host doesn't see us unloading the bikes from the tow truck. I also don't want the hosts to meet one of the most unpleasant men I have ever seen.

Once the bikes are out, I go to the driver's side door and hand the man my credit card. He is adding the tow to Onsted onto my bill when I notice he's typed in a total for over two thousand dollars.

"Excuse me, but that doesn't look right," I say.

"What do you mean?" he asks.

"Look at your iPad, man."

He looks, then begins violently laughing once more. "Whoops! I guess that means I don't get a tip now, huh?"

"I don't have any cash, man. You have a nice day," I tell him.

The man and his girlfriend leave and then I get a text from our host John.

"We will be home in a couple minutes," it reads.

Now that we're comfortably inside the gated community, I turn to my left and see a lake with a public boat landing. The wind connects with the water, creating waves that sparkle when the sun touches down.

Soon John arrives and puts our bikes in his garage. His wife Maya and their son Rex join us in the kitchen after they change clothes. When Sam tries to shake Rex's hand, Rex scoffs and bristles in Sam's face before leaving the room.

"Wow, okay," Sam says.

"What the fuck is that guy's problem?" I whisper to Sam.

"I have no idea."

"I thought you were going to beat his ass."

"He was pretty rude, wasn't he?"

"Fuck him. Let's just move on," I insist.

John is opening his liquor cabinet when we walk into the main room. "So do you guys want to kayak?" he asks.

"I'm good on the kayaking," I say.

"I'd like to go. I like trying new things," Sam mentions.

While he and John head outside, I peacefully sit in a rocking chair and watch them walk over to the lake. There is a certain peace to watching them happily load the kayaks into the lake. I massage the glass of wine in my hand and take a moment to appreciate the light that's creeping in through every window of the sunroom.

At dinner Maya and Rex enlighten us on their plan to circumnavigate the United States next year, which is surprising because I assumed the most exercise Rex ever gets is when he is smacked in the face by a dodgeball during gym class.

"Bikes riding okay?" John asks in between bites of a hot dog.

"Mine is actually quite fucked," I seethe. "Probably going to have to get it looked at tomorrow."

"What do you mean?"

"The chain keeps popping out of place when I shift. I don't know what's wrong. I've been to two different shops, and neither has fixed the problem."

"Do those people just not know what the hell they're doing?" John asks.

"I've had the exact same thoughts. It's like, Jesus Christ, all I want is for my bike to be able to cycle through the gears. But it's whatever. Hopefully tomorrow I can get it fixed."

"You know, I don't know much about bikes, but I can take a look at it, if you'd like," John offers.

"Feel free. I'll take any help I can get."

I munch down a few more hot dogs and go outside with John to take a look at the back derailleur, not expecting much. He fiddles around for a few minutes and then calls me over.

"Again, I don't know much about bikes," he says, "but you see this screw down here?"

"I see it."

"This screw is loose. It could be why your chain keeps falling off."

"That actually makes sense. I can't believe those bike shop guys missed that."

"I'm going to tighten it and put some Gorilla Glue on it. We will let it dry, and then let's try it in the morning. But, I mean, I think you should be shifting fine."

"You know more about bikes than you think you do, John," I tell the man.

I want John to be right, not just so my bike runs smoothly, but also because I want those morons in Kentwood to be shown up by a middle-aged man with more decency than those clowns.

"You guys want to watch TV?" John then asks.

"Are there any sports on right now?"

"For sure. Let's go sit down."

The first thing John turns on is IndyCar racing. He's such a fanatic that he tells us all about a sport that is so boring, you couldn't pay me to watch it.

"Just look at the way the cars go around the track. It's amazing," he says.

"It's okay, if you're into that kind of thing," I tell him.

He then pauses the TV to further explain his enjoyment. When John does this for a fourth time, I interject.

"Hopefully you don't pause this much during the Cavs game," I casually mention, trying my best to not be rude but at the same time express the simple fact that if this happens during a playoff basketball game, I will send him the most passive-aggressive stares a Minnesotan like myself can muster.

"Of course I won't pause during the game tonight. I just want you to admire how the cars move and the feel of the track. It's crazy."

I politely laugh, not sure the feel of the track can be appreciated from a sofa.

"Dad, you didn't do my laundry!" Rex yells when he comes down the stairs.

"Rex, it's not my job to do your laundry," John yells back.

"Yes, it is!" Rex says, then stomps back up the stairs.

"That's it! I'm blocking the Wi-Fi, Rex!" John yells after his son, but I don't think he heard him.

"How old is your son?" I ask, thinking Rex might have just hit puberty.

"He's twenty," John replies nonchalantly.

I nearly spit out my drink. Sam and I look at each other and then try not to laugh.

"It's perfect," John continues. "If Rex misbehaves, then I just shut off his Wi-Fi."

"And he's twenty, just to confirm?" I ask.

"Yes. He's twenty."

"I just wanted to make sure you said he was twenty years old."

"I did just say he was twenty years old, Q."

"Got it. So he's twenty years old."

Later that night, as the Cavs begin to increase their lead in the third quarter, a calm evening seems likely. John shares the less-than-captivating stories of his past, then transitions into explaining how instead of going to college Rex works at a grocery store down the road.

"He just doesn't like school. He only plays video games all day," John laments.

"My generation sucks. I'll admit that," I quip.

During the fourth quarter of the game, the Cavs lead slowly whittles away, as does Sam's phone battery.

"Do you have a charger I can use?" Sam asks John.

"Apple or Samsung?"

"Samsung."

"I have this one from Apple. It's way better than Samsung."

"I actually read that Samsung was better," Sam mentions.

"No, it's not. Apple is," John proclaims more aggressively.

"I don't know, I think—"

"No. It is. I know for a fact it is."

John then lists off stats to further enhance his point. I sit and marvel at the two go back and forth over something so unimportant.

"Guys, does it really matter?" I ask. "It's a phone charger."

John then stands up and walks to the kitchen to grab a charger, the brand of which I don't care to know.

The Cavs eventually lose after the Celtics Avery Bradley gets a lucky bounce and sinks a three-pointer at the buzzer. This forces the far superior Cavs to play at least one more game before they go to the finals. Like any day, you just never know how the tide will turn.

CHAPTER 16

"Try shifting gears. Just ride your bike around on the street for a little bit," John says to me from the steps of his front door the next morning.

I begin pedaling around the gated community, unsure how the bike is going to respond, but after a few times shifting through the gears, everything appears to be operationally sound. Not wanting to get too excited, I cycle through the gears again for the next few minutes.

True to form, John's theory about the loose screw on the derailleur has proven correct.

"I really don't know what to say. I think you've just saved my trip," I tell John, shaking his hand with extra gratitude. "Seriously, I don't know how to thank you."

"I really want a copy of your book," he says.

"I'll send you one."

Sam and I then leave, the sun shining on our last day in Michigan. We pull into a restaurant connected to a gas station around lunchtime. Chili dogs and pizza slices are on special, and the way my stomach rumbles, I can't wait to dive in.

"I hear you boys are going all the way to Maine," a younger man says on his way out of the restaurant.

My mouth is full, so I can only smile and nod my head.

"Yes, sir!" Sam belts.

"That's quite impressive. Best of luck to you, boys," the man says before getting in his car and driving off.

"Nice guy. You about ready to go?" I ask Sam.

"Let's *git* it on," he says with excitement, as if a plate of the finest chislik in South Dakota has just made its way into his stomach.

It's then, not yet out of Michigan, when disaster asks me for a dance.

A set of railroad tracks approach, and the road gets tighter. Sam buzzes ahead, and I'm still laughing at some of the banter from lunch. I look to my left, and a woman tries to peel out of a McDonald's parking lot, only to be halted by traffic. My gaze then returns to the road and the upcoming set of railroad tracks.

They are set at an odd angle, and the closer I get, the more my brain struggles to anticipate how to react. When my front tire grazes the middle of the tracks, I'm suddenly in trouble because the tire becomes stuck in the narrow metal.

I twist my handlebars to try to bust out of the divot, but my front wheel jams and gets stuck even further inside the groove. I quickly try to swerve out, but just as quickly my body is flipping over the handlebars. Momentum can't be stopped. I am going to fall off my bike.

Luck is not on my side. I'm not falling forward or to the right, either of which would be a relatively safe landing spot. Instead, I am toppling over the bars to the left.

"Fuck me," I yelp.

I try to put my left foot down to brace the fall but can't find footing. My groin connects with the handlebar, and this is not going to be pretty. The muscles in my left arm brace for impact as I fall to the ground. With this descent comes a loud horn and the squeal of slammed brakes.

Crunch.

I'm on the pavement and can hear the rolling of tires. I look up, and my face is two inches from one of eighteen wheels that eventually squeeze by.

"Oh my god," I whisper, unable to move as paralysis engulfs my arms and legs. My head swivels each direction, unsure if this is actually happening because it's all a blur. I look ahead, and the same woman in the McDonald's parking lot is staring straight at me, screaming into her phone.

"Get the fuck out of the road, Q!" a distant voice screams.

THE LONG ROAD EAST

"Q, get out of the fucking road!" the voice yells again, but I can only sit patiently with my hands on my knees.

Suddenly a pair of hands are lifting me up, and reality begins to return.

"Sam," I say softly. "I'm good, bro. I'm good. Just give me a minute."

Sam lifts me halfway to my feet and then I go the rest. He then picks up my bike and begins spinning all the wheels to make sure they aren't damaged.

The moment now feels like a gift, so I crane my neck and look up into the sky. Twenty seconds ago, I wasn't guaranteed to see the clouds ever again. It's taken me this long to appreciate what I could always see.

"Your bike is good," Sam says, holding my handlebars out for me to grab.

"Holy shit, what just happened?"

"Fortunately nothing," Sam says. "I did not want to have to make that call to your mom."

"I'm glad you didn't have to," I tell him, feeling a sense of appreciation for Sam's selflessness during the last thirty seconds.

The craziness of the moment elicits a laugh that drags on for what feels like an eternity, the laughter drowning out the morbid thoughts in my head. Death was inches away, but it's still hard to reconcile just what that means. I need to get back on my bike and keep riding, to repress what just happened before fear makes me want to never sit on a bike again.

Before this, potential risks were outlined by people who would never take a trip of this magnitude, but right now a spot next to the speculative armchair quarterbacks seems pretty cozy.

Sam tries to extract the positives from the situation for the next ten minutes before feeling comfortable enough with my disposition to forge ahead, but I'm not cured.

My mind replays the moment a million times over, the whole time wondering if there was something I could have done to avoid this situation. I would have pitied Sam if he had to call my parents. The way my mom would have reacted would have been unbearable.

Yet in many ways I wish that semi had run me over, ending my life at the ripe age of twenty-four. I'm not depressed or suicidal, even though it's cliché to say so, like an alcoholic who labels their drinking habits recreational. But I'm also not the happiest man in the world. I would love to get laid more and have more money to my name, but those are all things that can change, and most likely at the same time if things bounce my way.

More than anything, I merely think I've lived a good life. Going out now, on top and doing what I love, sounds immaculate. I'd be spared the ardor of testicular cancer or multiple sclerosis, a wheel pinning my head between the pavement providing an instant death. People would have talked about me with such joy, remarking only on my positive qualities, inclined to forget all my transgressions. I would have been a writing talent never fulfilled. Potential is so much sexier than failure.

"Excuse me, sir. Are we in Ohio?" Sam hollers at a kid who just started puberty once we later convene.

The boy raises his hand and opens his mouth with something to say, but by then we are too far gone to hear his words. A few strokes of the pedals later, a sign appears, welcoming us to Toledo.

At first glance, the city is a mess. Outside of the university campus, all the other buildings are decrepit and desperately in need of a facelift. People hang out on the sidewalks like they have nothing better to do, and they stare at Sam and I with such curiosity.

"Head on a swivel, bro," I tell Sam while we wait at a stoplight.

We aren't far from the destination. Across the bridge and a left turn later, loud banging emits from a hospital under construction. Only the hospital dons orange construction tape, even though the entire city feels in the process of something.

"I think we are here," Sam says when we pull up to a modest home soon after.

I go knock on the front door. No one comes to the door, so I knock again. I can hear footsteps shuffling toward the door, then stopping.

A lock is released, and the door slowly cracks open.

"Hello," says a man with glasses, half his face hiding behind the other side of the door.

"How is it going today, sir? I'm Q and this is Sam right behind me. Are you Jake?" I ask.

Surely this isn't Jake because the guy's expression indicates he has no idea why we are on the front steps. A short interval of silence follows.

"This is 127 Arnold Street, right?" I ask again.

"Jake said you guys wouldn't be here until around dinner," the man finally offers.

"We got in a little early today. Is it okay if we come in?"

The man looks back inside, and then back at me, like he hasn't yet put away something criminal. "Yeah. Go to the back. You can put your stuff in the garage."

We put our bikes inside and then are led to the upstairs of the home.

"You want to shower first?" I ask Sam.

"Sure," he says.

"Great. I have to call your mom."

"Huh? Why are you calling my mom?"

"Just to talk. Don't get all defensive."

"You're weird, kid."

"Dude, I'm not going to pull anything sketchy," I try to assure Sam.

"That's a first."

When I call Sam's mom a few minutes later to explain the near-disaster that almost occurred, she responds sympathetically.

"You should call your mother. She would like to know," she advises once our conversation has run its course.

I am spread out on a mattress looking out the window. The sun is still shining, creating a peaceful Midwestern spring day that feels so earned after the winter I had.

"I don't know what I'd say to my mom," I sigh.

My mom was never a great sympathizer. I vividly remember as a young boy coming home one night from a family function. She told me multiple times to brush my teeth, but I kept refusing. When she had enough of my insolence, she grabbed me by the throat and slammed my head into the shower door.

"Brush your fucking teeth, Quentin!" she yelled that night.

The scariest part of that ordeal was I could tell my mom was tired of me. She wasn't going to give me up for adoption, but she didn't want to be my mom anymore that day. I'm sure every parent has those days, but I couldn't comprehend this as a young boy with tears streaming down my face. I could only assume my mom no longer loved me.

I go against Sam's mom and neglect to share the information. I don't want to bring up a reality that cannot be altered.

After showering, Sam and I walk around Toledo looking for a place to eat, eventually settling on a fast food joint that serves seafood. Construction workers begin to file in after 5:00 PM. I know I should make a phone call to my family, but instead I post what happened on Facebook so random people take the time out of their day to give me attention.

"You finished?" Sam asks, grabbing his tray.

"Yeah. Let's head back."

When we get back to Jake's place, he is sitting in the kitchen. We shake hands, and I take a step back to allow him to get out of his seat comfortably.

"You guys hungry, thirsty?" he asks.

"We just ate, unfortunately," I say.

"But are you thirsty?"

"I'll have a beer with you," Sam pipes in.

The three of us drive to a Mexican restaurant. Together we devour the complimentary chips and salsa every time the server brings a new bowl out, and Sam and Jake drink margaritas like they are old buddies while I scan the room for wandering female eyes. Small talk is not my medium.

"Are we thinking about another round?" the server asks after Jake finishes a second margarita.

"I think we're good," Jake answers.

"Is the bill together or separate?" the server then asks.

"All on one," Jake requests, handing the server his card.

"Wow, thanks, Jake!" a delighted Sam bellows from the depths of his satiated stomach with so much happiness, it makes one wonder why the little things in life aren't more appreciated.

"Let's get out here. I think we all have big days tomorrow," Jake then says.

CHAPTER 17

When I think of what makes a ride difficult, it's never the physical pain. I can get over a sore calf or an itch that's formed underneath my left thigh. I can even persevere through the gusting winds that sweep across the plains of the Midwest each spring.

Instead, what really makes a ride so difficult is the monotony of it all. Most days are spent stuck inside my own head, trying to forgive myself for all the idiotic mistakes I've made. The past is never easy to let go of, but it also shouldn't be harbored like a tugboat docked at port.

No matter how much I try to assure myself that I'm merely human and that perfection is unattainable, my mind still doesn't believe me. The only way I exit these draining headspaces is when I finally catch up with Sam, or the road becomes so narrow that I have to focus all my efforts on riding in a straight line.

But those moments tick off faster than the last ten minutes before bar close, and then it's back to the cruel self-punishment I inflict upon myself.

This trip has followed the same pattern: motivated beginning, eat lunch, get bored, fantasize about having sex with my ex before complaining to Sam that the day is getting long. It's all capped off by digging deep to find new motivation for the last ten miles.

Right on cue, the day has reached its penultimate juncture.

"We have about ten miles left," Sam says before we face one more long stretch that's surrounded by hordes of corn.

"I guess that doesn't sound too bad," I reply, but then the next ten turn out to be brutal, a headwind gusting in my face and making me wish I was anywhere but on this bike.

THE LONG ROAD EAST

The miles don't feel like they will ever end until the Huron, Ohio, water tower comes into view. The adjacent highway is about two thousand feet to the left, the speeding cars barely audible from my position. I always count on the highway to be a reminder that we are moving forward and not further into the depths of rural purgatory.

When I finally catch up with Sam, he is parked on the side of the road looking as peaceful as the times when he has a pizza box in his lap.

"Take a break, young man," Sam tells me, so I take out a bag of gummy worms.

"Man, these taste good," I say, caressing each worm before sticking the more colorful end into my mouth.

"Okay, porky. Enjoy those gummy worms," Sam chirps.

"Fuck off. I'm still ripped. How far until we are there?"

"Ripped is an interesting word choice, but to answer your question, we're about five miles out."

"Five miles? What the hell are you good for?"

"Dragging your sorry ass across the country," Sam quips, then pretending to throw a few punches to my stomach.

A few mountainous turns later we are on a sandy road next to a slow-moving river. The water is at peace, much like the retired people who inhabit this trailer park. An old guy with a beer in his hand and meat on the grill nods with a welcome-to-heaven crane of his neck.

A few trailers later we stop and meet our Warmshowers host, a bearded man named Wally.

"You guys want a beer?" Wally asks after we put our bikes in his shed.

"I'll have a beer with you," Sam predictably muses.

"Do you have any wine, Wally?" I ask.

"I do. It's in the fridge."

A glass of wine feels like a fair reward after today's grind. I am so exhausted, and with my birthday two days away, a few glasses of Pinot Grigio seem appropriate.

Sam and Wally begin downing Miller Lite with the same pace that an elephant drinks water, and before the full-course meal is out,

I can tell sobriety and a night of intellectual conversation is not on the horizon.

This is why I hate Sam sometimes. He's so good with people, a chameleon that can turn into any mold a host might see fit. Everyone we have thus far met has liked him. His gregarious nature detracts from my ability to showcase my own talents.

Not that it really matters. With Wally being a former veteran who has gone through divorce, he and I don't have much in common. I'm okay with that, because to pretend that we do would only be more emotionally exhausting.

After dinner, Wally's late-life lover Eleanor joins us as the food comas begin to sink in. Her daughter comes through the screen door a few minutes later, and my relaxed state gives way to indecent thoughts.

"Ask them if they need anything before you go to the store," Eleanor instructs her daughter.

"I could use some Listerine, if you're offering," I say.

"Geez, kid!" Sam interjects.

"What? I'm almost out."

"Sam, it's fine," Wally interrupts, waving his hand and further cementing the bromance where the Miller Lite has not already.

Elanor turns to me after handing her daughter money. "Do you have a girlfriend, Quentin?" she asks.

"No," I tell her. "I used to, but it didn't work out."

"Oh, I'm sorry. Why not?"

"A number of reasons, but I wanted to sleep with other women, and obviously that didn't sit well with her."

Elanor's face drops. In her mind, I have instantly gone from being a charming young man to a plague on society. Her eyes spew hatred, like every other woman from the last few months who didn't appreciate the truth I so inconveniently offered to them.

It upsets me that Eleanor expects me to fit her mold. If her facial expressions were less obvious, I might give her a pass, but she's looking at me like I've just committed a crime.

"I see," Eleanor mumbles through a mouth that's half ajar, then stands up to go clean the dishes.

THE LONG ROAD EAST

I am a quarter of the way into the next glass of Pinot when my eyes slip shut. We will be in Cleveland the next day for my birthday, but I no longer feel like celebrating.

CHAPTER 18

Rain persists under gray skies. It's only worsened as we inch closer toward Cleveland.

"This is so fucked!" Sam laments when we pull over at a gas station for lunch.

"What's wrong?"

"These cars. They don't move over for shit!"

We have been on an alternate highway that takes us directly into Cleveland, and despite there being two lanes, cars whizz past with no regard for our safety.

Usually cars are within arm's length, but that's on country roads. Now we're in the suburbs of a bigger city, on a busy road that has two lanes going with us. It's not inconceivable to expect cars to shift a lane over when they see us. Most unnerving is that despite our lights and bright-colored clothing, cars might not be seeing us underneath the clouds and rain.

"I'm seriously getting pissed," Sam continues to bemoan after I come out of the bathroom-less gas station with chicken and mashed potatoes.

It's the first time all trip I've seen Sam genuinely upset about the conditions of the road.

"All we can do is power through this," I tell him.

"Yeah, I know. You don't have to remind me."

"Well then get that shitty look off your face and get us to Cleveland. It's almost my birthday."

"Just eat your food, Quen-ton."

"I'm guessing you're pissed?" I ask.

"No. It's just that if I wanted your lip I'd go in your pants and get it."

THE LONG ROAD EAST

"Like I haven't heard that one before, Sammy Boy."

When we resume riding, Sam begins to distance himself, and I'm alone in enduring the cold splashes of water that gets kicked up by the cars. Over time, grimy water works its way up the tires of my bike and onto my back.

The shoulder keeps narrowing the further I ride. I have become a bigger nuisance to the cars because they have to slowly pass to avoid broaching the other lane or hitting me. I finally find a sidewalk, but that is awful as well because there are cracks every five feet that when combined with the rain become another inescapable irritant.

For miles I oscillate between both path and road, hyperaware of being in the road but also hating every block of the sidewalk. I haven't seen Sam for almost two hours and don't have any missed calls when I pull over to stretch and check for new Tinder matches.

"He can't be far off," I tell myself. Google Maps has me a few miles away from the destination, yet there is still no sight of Sam.

I take one final stretch under another onslaught of rain, and when I look up from checking my phone Sam is just up ahead. We meet and then ride into a bourgeois sector of Cleveland.

"What if we got to stay in one of these places?" I dream as we ride out the string together. "This only has to continue for the next few miles."

"I wouldn't complain," Sam says.

Predictably, the housing situation turns more middle-class by the time we come to a stoplight. Everywhere we stay is more than satisfactory, but I keep building these romantic notions in my imagination, like if we stay at an upscale home we will be treated to a night on Lake Huron in a yacht filled with strippers and expensive cigars.

"Oh, fuck, man," I say to Sam as we wait for the light to change.

"What's up?"

"It's embarrassing to admit this, but I've been thinking about her all day. I don't know why. It has to be the rain that puts me in this funk."

"Dude, you have to get over your ex."

Sam looks my way, and I want to tell him it's not that easy.

"So then what are you thinking about?" he asks.

"I don't know. Just how she is with someone else, how her and I will never have a family together. You know, the depressing shit."

"I wouldn't worry about it, Q."

"Why not?"

"She'll probably blimp out," Sam suggests.

I laugh. "Anything is possible, I suppose."

We arrive at the Warmshowers early, so we sit on our host Carrie's front steps while waiting for her to get home. The rain is a crazy beast. Simultaneously wet and dry, it comes and goes as it pleases, in no mood to compromise with cyclists.

My relationship with the precipitation is complicated though. I miss the rain when I sit inside at night; the sound it makes when connecting with my jacket or face. Sometimes it hurts my forehead, and other times it feels like I'm being baptized. The volume ebbs and flows, the decibels ranging from sleepwalking to headache-inducing. The rain and I aren't friends, but we aren't enemies, like De Niro and Grodin in *Midnight Run*. I'm infringing on its objective, yet it still finds time to forgive my trespasses.

Carrie pulls up in her sedan and parks it outside the garage. "Take as much of your clothes off as you can before you come inside," she tells us.

While removing my pants from my legs, Carrie's dog comes out and licks every strand of hair on my knee. "Could you not?" I ask the beast, but it only looks at me in confusion.

Upstairs in our room, there is a gray cat sitting on the couch. It merely bats an eye when I set my things down a few feet away.

"Shoo. Come on, move," I tell the feline, but it won't budge, instead visually inspecting every inch of my body with its beady eyes.

"Just go downstairs," Sam says. "Worry about your stuff later."

I do just that.

"You guys want a drink at all?" Carrie asks as Sam and I huddle in her kitchen for cheese and crackers.

"I'll have a beer with you," says a voice that's not mine.

Carrie then hands Sam a key as I gobble down enough white cheddar to feed a book club.

"Take this key when you leave. Just make sure when you come back tonight that you lock the door," she says.

An hour later, Sam and I order an Uber that takes us downtown to a sports bar with a Cleveland Indians game on every TV.

"Grey Goose Red Bull, please," I tell the bartender, but nine dollars later my palette has not been cleansed.

"You want to go to another place?" I ask Sam.

"Yeah. Let's check some other places out."

We walk down a softly lit street to a nicer bar with a wooden interior. With not much to say, Sam and I only dabble in small talk. We are trying to celebrate my birthday but are failing epically. This night needs a spark. I don't want to spend my birthday talking to Sam about the ergonomics of cycling and the unpredictable futures that await us back in Minnesota.

When we head over to a third bar, the night seems destined for mediocrity. I'm not a birthday guy, but this is my golden one and thus feel inclined to make it special even though it feels the same as any other May twenty-fifth.

"I'll take a Long Island, *without* tequila," I emphasize to the burly bartender.

"We don't use tequila. We use a mix," he says.

"Okay, that's fine."

Like a winning lottery ticket, lightning then strikes.

"We should go to the strip club tonight," Sam proclaims, a smile slowly appearing on his face.

"Are you sure?" I ask, quickly thinking of the financial repercussions.

"I'm going to the ATM to get some cash, and I'm giving you a hundred dollars."

Sam goes over to the machine, and I grab the attention of the bartender.

"Sir, my buddy and I are trying to find a strip club. Which one should we go to?" I ask.

"They're all fun," is all he says.

Sam comes back and hands me a blue one-hundred-dollar bill. "Did the bartender say where we should go?"

"He said they're all good. Let's just Google one."

Sam finds one and then the Uber soon arrives. A large friendly black woman with a loud voice is behind the wheel.

"You guys are going to the strip club tonight, eh?" she asks with an inviting twang.

"It's my birthday, so we gotta do it," I tell her.

A song revolving around a woman asking for a man to put his body on hers plays through the speakers of the car, offering an apropos introduction to the rest of the evening.

"Can you turn it up?" I ask the woman.

"I got you," she says.

I press my face against the warm glass and look out the window while Sam flirts with the driver. We're out here, and right now it feels like a dream.

The car methodically brings us out of Cleveland and into an industrial neighborhood across the river. A mix of colors beaming off the strip club represent the only lights in the vicinity.

"We'll tip you on the app," I tell the woman, and then shut the door.

After getting past security and up to the bar, I order two Long Islands, which account for a quarter of the money Sam gave me. I hand him his drink and then we find a table, allowing me to melt into my seat and enjoy all the visuals the club has to offer.

Perhaps I'm an immoral cancer for saying this, but I love strip clubs. There is a certain beauty to them, no matter how morally compromising they may be.

Tonight won't be easy. It would be foolhardy to think I could walk in here and have chicks drool over me. Vanity only goes so far in a place where women work to mask a pain that exists in their lives, and men follow just to stroke their faltering egos.

Sam has a huge grin on his face.

"You've been to a strip club before, right?" I ask him, careful to point out the potential for a young man to get lost in his vices.

"Of course I have!" he claims, rattling off a city in the middle of Iowa that was the site of his first foray.

The first woman to entice me into a back room has dark black hair and a pseudonym that isn't Candy or Roxy.

"You are just so cute," she says with a smile, sliding her left leg along my hip and grabbing the back of my head with her free hand.

Her hips pump back and forth. I want to mention her actions aren't arousing, but voicing my displeasure seems rude.

"This is so awesome," I exaggerate as a song nears completion, marking the end of our overpriced ritual.

"Do you want to keep going?" she asks after the four minutes conclude.

"I'm good. Thank you though," I politely reply.

Sam is gone when I come back out. I scan the club's expansive layout and don't see him. My guess is he is burying his beard in D-cups behind a curtain somewhere.

I peer around the room. A Russian man wearing a suit is flashing money as his bottle of champagne sits comfortably in chunks of ice. I suddenly want what that guy has. I want women draping on my arms in pursuit of my money. I want to feel like money has no bearing on my decisions. I don't want to have to count pennies and have that impact how long I'm in the back room. With where I am at in life though, that fantasy simply has to wait.

Occasionally throughout the night I flirt with proactive women who slyly run their hands along the parts of my body not covered by clothes. Happiness is not going to find me in a place like this. I'm better off finding Mrs. Right and fertilizing her eggs. She would bring everlasting joy, and I could eventually die an old man feeling like I contributed something to the world.

But as I sit in a chair ripe with premature ejaculation, I don't want any of that. I want instant gratification, sex, and all the feelings of potentially hating myself that the next day will bring. I want to wake up in a few weeks and feel an irritating tingle streaming out of my anatomy. I want to then stress out about it, lie to myself that I am done with promiscuity and soon after find myself falling victim to my own insecurities once more.

Right now, I am consciously recognizing my own duality, letting the evil win out and demolish all the ethical principles my parents worked so hard to instill in me as a child.

After a few more thoughts revolving around self-doubt, a young woman gets on stage, and our eyes instantly meet. She doesn't have the most attractive face in the world, but her body is otherworldly. Her legs are long, accentuated by the dark black heels strapped to her feet. Her stomach is tight, small muscles peeking out the sides of her midsection. Her eyes rarely leave mine as she dances. It's all a ploy, her way of reeling me in once her set ends.

"Everyone give a hand for Charlotte," the DJ says after a few songs.

I stop slouching and sit up in my chair. After a few songs, Charlotte elegantly steps down the back of the stage and wades through the empty red chairs that have been vacant all night. The closer she gets the more my hand reaches for my wallet. More tingling sensations form just above my right hip.

"Want a dance?" she queries, running her hand along my arm, practically lifting me out of my seat.

"How could I say no?" I chuckle as I'm already being led into the back room.

She holds my hand and then brings me down to a grainy cushion, releasing my hand once I've sat down so she can close the curtain behind us. I put my left hand on her ass, and she takes it well; but then our eye contact vanishes, and everything that seemed possible seconds earlier is all but extinct.

"You can't even fake it a little bit?" I ask.

"What do you mean?" she retorts defensively.

I open my mouth to speak, but no words come out. I'm drunk, and if there was a morsel of an opportunity for sex, it's more lost than that golden retriever in *Homeward Bound*.

We walk to the ATM after the song ends, and I am further humiliated. She hovers over my shoulder as I try to conceal the digits to my account, looking around the now-mostly-empty strip club with a roll of the eyes that can't believe I am this pathetic.

It's an eight-dollar charge from the machine just to give this woman money. The balance of my bank account pops up, and I quickly withdraw cash so we can both stop wasting each other's time. Emotionally empty, I walk back to my seat and wait for Sam.

Twenty minutes later, Sam comes out with a big grin on his face.

"The…the…Champagne room," he finally unearths, exasperated and out of breath as he joins me on the red chair idling next to me. "She's going to come out to Maine to meet us."

"Oh, fuck," I say.

"What?"

"How much did it cost you to fall in love?"

"Four hundred dollars."

"Four hundred dollars!" I wail, quickly calculating how much chocolate milk that money could buy.

When Sam says four hundred dollars, the freedom of biking cross-country feels threatened. I need another drink.

"Bar's closed," a woman with glasses says when I rest my elbows on the varnished wood.

I walk back over to Sam. "We might as well get out of here," I tell him as we stand in the smoking area outside.

"Probably a good idea," he agrees.

Back at the house I lay my head on a pillow and wonder how I became so consumed by women. Six months ago, I had a future bride, but now I'm chugging water and looking at the ceiling in disbelief. My life feels like an utter failure. *Happy birthday, young man.*

CHAPTER 19

The next morning, I thank a deity my head isn't pounding. Rolling over, it will be another blessing if my stomach doesn't flop.

"Get up. We have to get going," I groan to Sam.

I go into the bathroom and endure one of the more painful visits of my life. The cool tile feels good on my feet, but my sphincter is warped and the inside of my asshole burns. Slowly melting, I try to come to terms with the previous night.

Sam showed me a great birthday, impeded only by my incessant struggle for sex. I can't remember if he was joking about a stripper meeting us in Maine until he mentions it again, only this time with less gusto and a stronger sense of realism.

"You're nuts," I tell him.

I need an orange juice and something greasier than my hair to soak up my physical and emotional ailments, but the only vitamin C nearby is inside the multivitamin I take every morning.

"What did you guys do last night?" a chipper Carrie asks as she stirs her morning coffee.

"Strip club," I say.

"Oh, nice," she says, and it's hard to tell if she's bothered by the sentiment.

Following my not-so-tactful announcement, we retrieve our bikes from the garage and exit unceremoniously.

The relentless rain is pouring as a few minutes later we stand outside a brick building trying to forge a plan. The nausea of the previous night is slowly leaving. I got drunk, but not enough to be in dire straits for the next twelve hours.

THE LONG ROAD EAST

It's downhill most of the way out of Cleveland, so we don't have to work too hard to get where we want to go, which right now is ideally a sit-down place that serves high-calorie burgers.

We soon find ourselves pulled over at a BP gas station. I bring a chocolate milk to the counter and wait a couple minutes for the hippie working the register to fetch a couple slices of pizza from the rotating oven.

Back outside, a young man approaches Sam and I as we quietly eat our food.

"Hey, hey, how we doing, fellas?" he asks, introducing himself as Black.

"Still getting over last night," I answer.

Black finally stops twitching and is then able to say a few words. "You wouldn't happen to have a couple bucks, would you?" he asks. "I need gas to get to this house just over there to go pick up some money. I can pay you back if you want to wait for me to go get it."

Black is an enigma. He doesn't appear harmful, but I don't believe him when he says he is using the money for gas. I look at Sam for guidance, but he hasn't made up his mind either.

"I got you, bro," I tell Black, reaching into my wallet to grab a couple ones that did not make their way into a G-string.

I'm not eager to give him the money but doing so feels better than saying no. Sam gives him more money than me and then Black goes into the store.

"You think that guy's actually going to buy gas with that money? It seemed like he wanted money for drugs," I say to Sam.

"I don't care what he uses it for. He can go buy drugs if he wants to," Sam responds, then walks to the trash can to throw away his empty wrappers.

The alcohol slowly drifting out of my body, we steamroll out of Cleveland and into the suburbs, ending up on magnificent back roads filled with dark green trees. The rain makes for slick riding, and as we hurl down hills at breakneck speeds, calamity could occur if one of us loses control, but it doesn't, making this sequence some of the best riding of the entire trip.

Soft drops of rain lightly hit my forehead and then roll down my face. This speed continues for ten minutes until I am snapped back to reality at the bottom of a hill.

"Bro, that was one of the most fun times I've ever had riding a bike," I tell Sam, not trying to contain my elation.

"I got up to forty-four miles an hour," he replies, pointing at his GPS attached to his handlebars.

Water immediately finds its level when the next few miles flatten out. And then, as if we were going down a hill out of Cleveland, we have to climb Mount Everest to get to Chardon and meet our hosts.

Every stroke of the pedals is agonizing. I lean slightly over my handlebars and try to foster as much momentum as possible from my paltry abs so I can push up the hills. The weight from Joad is pulling me back down, and I want to unhook her and bid her farewell, but I told that girl from the start we'd go until death tore us apart.

Now is the moment I thought would eventually come, where we see a shift in terrain and realize our bikes are somewhere they've never been before.

Halfway up a grueling hill, we both have to get off and begin pushing our bikes. The rain elicits shivers with every breath that exposes the apple in my throat. A school bus that passed us on a steep incline a few turns back roars down.

It's slamming on the brakes, and I squint my eyes wondering if Sam's involved. Deep, heavy breaths emit from my mouth over the next thirty seconds until finally Sammy Boy appears at the crest of the hill.

"I thought you might have got hit," I tell him as my arms gyrate explosively and my body tries to stay warm.

We are only ten miles away, but they end up taking two hours to traverse.

It's then we come to the top of a house that's situated on a steep incline. We ride up the small bump just before the driveway, arriving at a house that's without an obvious front door. Letting intuition make the decision, I carefully walk a cement path leading to the door that appears to be the one most appropriate to knock on.

THE LONG ROAD EAST

A woman named Mary answers while I wipe beads of rain from my brow. She's cute for an old lady, her sandy brown hair having survived her younger, more vivacious years. She's not exactly enthralled to see me.

"Hi. I'm Quentin Super."

"You were supposed to give me a heads-up when you were close," she says with condescension.

"I know, but unfortunately my phone died, Mary."

"Yeah, yeah. Get your stuff and go through the garage. I'm actually in the middle of something right now."

Sam and I strip off as much of our wet clothes as we can while Mary lectures a large black dog on the finer points of sitting and staying. I quickly scamper upstairs and begin embracing the sizzling hot water of the shower. It is the final remedy for all the debauchery that occurred in the last twenty-four hours.

That evening we all sit down for dinner. Mary and her docile husband Gary have prepared a nice spread of meat and potatoes, but the best part of the meal are these little boxes of orange juice. I drink four of them before the citrus begins to dry out my tongue.

"You said it was your birthday last night?" Mary asks at one point in the meal.

"More or less. The big two-five," I joke.

"Did you celebrate?" Gary asks.

"We did."

"What did you do then?"

I can't keep my big mouth shut.

"We went to a strip club," I unapologetically inform him.

Gary and Mary find no appeal in my revelation, and Sam is looking at me with as much contempt as he effused during my desperate Tinder romps throughout Wisconsin.

"Oh," Mary replies.

"We just wanted to check out the downtown scene mainly," Sam says in hopes of saving face.

"I see," Gary says.

"So you guys Cavaliers fans?" I ask.

"Sort of," Gary chimes.

"Was it pretty wild when they won last year?"

"I guess," he says, his mind still coming to terms with the fact that no matter how long of a shower I took my actions cannot be erased from his thoughts.

"Did you go to the parade?" I question further.

Gary gives me the *are you serious?* look. "No. Too many people," he bristles.

He then leans over the table and buries his head in a plate of mashed potatoes.

"Are you guys going to watch the game tonight?" I then ask.

"We don't get cable," Gary murmurs while a small particle of potato hangs from his lip.

After dinner we move to the living room. Sam has a beer with Gary, and I pass out on the couch around eight thirty while a black-and-white movie plays on a small TV. It's my last day being twenty-four, but it won't be the last day I draw the ire of a middle-aged man.

CHAPTER 20

It's going to be a tough pull from across the Ohio state border and into Erie, Pennsylvania. The sky already has a *screw you* look to it. A few drops of rain seep through my helmet and onto my hair, and already I want to be back in Mary and Gary's shower scrubbing the sin off my body.

We first rest at the stop sign of a three-way intersection. The wet and disjointed gravel beneath my tires starts to suck me in when an old man nearly sideswipes me. He must have recently fallen out of his rocking chair because as he comes within centimeters of my left shoulder, I can see him looking in the side mirror to see how close he is.

This has been a common theme all trip. People have no idea how to drive next to cyclists.

"Is that old man fucking retarded?" I angrily ask Sam as the man's gold Buick slowly peels out into the lane of traffic.

Sam laughs as he too watches the vehicle creep into a right turn, the car still closer to the ditch than the median.

"People like that just shouldn't be on the fucking road," I cry.

"God, kid," Sam chuckles. "You're such an asshole."

"Seriously though. Why would he move closer to us? That isn't even logical. Some stupid fuck is going to hit us one of these days."

"Let's just get going. I can't stand around and listen to you bitch all day."

Rain and gloom persist as I try to shrug off the early nuisance. Sam further distances himself and then I begin to sift through dark feelings.

I can't stop thinking about my ex. The more miles that tick off, the more she becomes centered in my thoughts. I think about how in the weeks after we broke up, she asked for her clothes back, an unimportant yet memorable item being this gray shirt that was not long enough to fit my awkward body. But it's not what she took that stung.

Throughout our time together we exchanged little trinkets, and without realizing this, it was a sign that we were becoming closer. Most of the items were little in value, but the day she gave me back one of my senior photos from high school, I couldn't internalize my sadness.

That pocket photo, the one in which I had a weird haircut, acne, no facial hair, and wore an ugly blue-collared shirt, was not just a reminder of how unfashionable I used to be. Her returning that photo was also her giving back her acceptance of me: the good, bad, and unfashionably ugly.

"Please don't," I said at the time, fighting back tears.

She then took out a small red piece of cloth. I thought I was going to throw up. *Quentin's Best Friend* was embedded into the fabric.

"I don't want it anymore," she said while unsympathetically holding it in her hand.

"Please. It will break my heart if you give it back," I clamored.

She could have tossed it in a dumpster outside or burned it in a microwave. The fate of the whittled fabric didn't matter. I just didn't want her to give it back.

"I don't want it," she sighed, setting the fabric on a table and slamming another nail in the coffin of my battered heart.

There were a million things I could have done to avoid that moment, or any moment where I died inside each time she looked at me and wished I would be better than the narcissistic, hormonal person I was. She gave me a million chances, her able to see a part of me I couldn't even see in myself. And each time she tried to show me what I could be not only for her, but for myself, I rejected it, reverting back to a devilish trap I thought would make me happy.

Perhaps most upsetting about the relationship ending was how I implicated someone I love in my own struggles with self-worth. I wasted her time, time she could have spent finding someone with more to offer than insecurity issues and a hankering for Grey Goose.

And on this day, somewhere along the border between Ohio and Pennsylvania, I can't shake the thoughts of her. I think about

THE LONG ROAD EAST

a road trip we took. We were supposed to make it to Las Vegas and then California, but on the first day my car had problems.

It was the weirdest thing. We were in Nebraska and suddenly my car began to shake. I could hear coins jingling inside my glove box.

"That's the strangest sound I've ever heard," I said.

We pulled over and tried to diagnose the issue, which proved futile because neither of us had any idea how cars worked. It took ten minutes just to get the front hood opened.

"We are just going to have to keep driving," I said, looking out at the vast wasteland of Nebraska.

I had been familiar with desolate places, but this was different. We drove for two hours and didn't see a town with more than one thousand people.

Later that afternoon we finally passed a sign that said North Platte, Nebraska would be coming up in sixty miles. A quick Google search discovered the town had over twenty thousand people. Certainly, there would be an auto body shop there.

For the next hour I drove eighty miles an hour down this backroad interstate trying to reach a town I only hoped would have a service station. With every mile the shaking became more prominent, to the point where the whole left side of the car was rattling.

She then got on the phone with a guy who claimed to know a thing or two about cars.

"Your wheel might be loose," he said. "I would stop driving right now because it could come off and start rolling down the highway."

Not wanting to stop, I kept looking at my GPS that was propped in between two cup holders. The miles slowly kept dwindling—thirty, twenty, ten. But the noise also kept getting louder.

When I looked over at her, I noticed she had begun crying.

"What's wrong?" I asked.

"I wanted to see the ocean with you," she whispered through tears that acknowledged we weren't going to make it to the West Coast.

"It's going to be okay. I promise I will take you to see the ocean someday."

I never did honor those words.

We eventually pulled into the North Platte city limits, only to stop at a repair shop that wasn't open.

"Let's just get to the motel," I said.

We only needed to get three miles and then I could deal with the car in the morning. Getting there was more stressful than bounding down the interstate. We pulled into the downtown, and it was full of construction. I thought about the civic responsibility I had to pull over and just eat the cost of a tow, but at the time I was on minimum wage and willing to do anything to save money.

"Holy fuck!" I screamed. "The wheel is like *this* close to coming off," I declared while looking in the side mirror and seeing the wheel dangling by a thread.

The hotel was on the right, and then a red light came. My Nissan was pleading for mercy. I wanted to give the car a hug because when you forget the obvious fact that a car is an inanimate object with no feelings, you start to realize just how much of your life that car has been there for.

The light finally switched, and we turned right into the parking lot. What started as a day filled with dreams of LA beaches was soon dashed for the confines of a Best Western in rural Nebraska.

"It'll all be okay," I assured her after we settled into the room.

Even if the car was somehow totaled, it didn't matter. Going through this stressful situation had me convinced I could spend the rest of my life with this woman.

Only a fool would have let something that good get away.

And all this comes back to me as I pedal toward Erie. Nothing has changed even though we are past the border. Sam's phone slips from his hand as he scrolls for directions. It takes a leap into the middle of the highway, landing facedown in the road. One car passes over without smashing it; then Sam scurries onto the highway to retrieve the lifeline and restore our sanity.

One more long stretch to go. Sam jumps ahead, and I revert back to somber feelings. I get off my bike at one point and put my hands on my knees. I'm not tired, but I have to stop and think about how abhorrently selfish I am. I gave up my relationship to chase women, but they have been few and far between.

I gave up on reality for a fantasy.

Even worse, what our relationship meant will lose clarity with the passing of time. Soon I'll forget all about my ex and the turmoil I endured in the months after our split. Life will begin again, and her memory will vanish. Even though I will always know who she was, as I become older I won't know anything more. Her evolution will be experienced and appreciated by someone other than me.

When I once again stop to stretch and rid my mind of harbored negativity, it becomes clear that recovery isn't imminent.

I grab my sleeve and wipe a tear off my face. It would be so nice to talk with someone, to tell them I'm not a horrible person, and that the bad things I did were merely lapses in judgment. But even that is a foolhardy wish. The emotional cavalry is not coming. No one cares about my problems, and nor should they.

"How are we doing?" Sam asks when we gather ten miles outside Erie.

"Not good," I concede.

"What's going on?"

"I've been to some dark places in the last few hours. I feel so guilty right now."

Sam nods his head. "The day's almost over," he says.

I look again into the sky. I hope this pain soon ends.

CHAPTER 21

Sam and I spend the next two days in Erie, Pennsylvania, with an older couple who have a dog named Bud. Within an hour of being at their house I have drank an entire half-gallon of chocolate milk.

"Oh, that's okay," an older bird by the name of Sue insists after I inform her of my piggish display. "Craig can go buy more. Please, eat anything you can find."

I'm later sitting on La-Z-Boy when in walks an Asian kid named Jake. Wearing glasses and a look of uncertainty, he looks lost, and I don't like him because he's following my lead.

When Jake finds out Sam and I are spending two nights, he wants to stay an extra night as well. And after telling Craig I need to go to a bike shop, Jake is suddenly also keen on going.

"Jake, you know the wind is going to be in your face all day once you get to the Midwest, right?" I tell him.

"I know," he confidently replies.

"How are you planning to overcome that?"

"I'll be fine," he assures me, citing no evidence other than his unwavering self-belief.

Jake is frustrating because we share many of the same outlooks on life.

"That guy isn't going to make it," I tell Sam, hoping he will see the same flaws in Jake that I do.

"Just relax, Q. He's a nice guy."

"Is being a nice guy suddenly a prerequisite to getting people to believe in you?"

Sam smiles. "Bro, you have so much anger inside of you."

THE LONG ROAD EAST

At the bike shop, Jake debates buying more gear. He only has two water bottles, so I urge him to buy a water bladder. After he rejects my advice, I walk over to Craig.

"Craig, you should tell him to buy a water bladder. He is going to need it out west," I tell the elder statesman.

"I'll talk to him," Craig responds, but Jake still refuses to buy the water bladder.

I throw my hands up and exit the store.

"Everything okay?" Craig asks a few minutes later.

"Yeah, I'm fine. Any chance you could take me to get a haircut?"

"We can do that. I know a guy," Craig says.

Craig drops Jake off at a mall and then takes Sam and I to a local barbershop. It's a family-owned place, a father-daughter tandem.

"You ready?" the older man asks me. I sit down in his swiveling chair. He does a stellar job on my sides and then asks about my beard.

"You want me to match it to your hairline?" he asks.

"Sure," I respond, not knowing exactly what that entails.

When he finishes, I look like an ugly young boy, a few specks of hair on my chin and virtually none on my jawline.

"Do you like it?" the man asks.

"Yes. Thank you," I lie.

After paying I saunter over to Craig.

"Is something bothering you?" he asks while we sit and stare at Sam get his hair done.

I am in a shitty mood but don't want to admit it. "He took off too much of my beard," I discreetly whisper so the barber doesn't hear me.

"So? Is that a problem?" Craig asks.

"I feel self-conscious."

Craig laughs in a way only old men who know petty things like this don't matter can. "You'll be alright," he says, and once Sam is finished, we head back to the house.

That night we all go to a minor league baseball game with a few members of Craig's family. Sam is already a couple beers in when he decides to leave his seat to go hit on Sue's sixty-year-old sister.

"Don't do anything stupid," I yell after him.

"Shut up," he says over his shoulder.

This leaves Jake and I to take in the ball game and admire the grace these major league hopefuls play with.

"You like baseball, Jake?" I ask, reaching for my soda.

"No, not really. I'm Asian," he jokes.

"Fair enough. I'm not a huge fan either."

"You like women though, right?" I then ask.

"Definitely."

"You sure? You look like you swing both ways," I quip.

Jake laughs just enough to make me think what I was said was funny.

"No," he then says. "I only like women."

"I know. I was just giving you shit."

"But I did get laid one of my first nights on the trip."

"Dude, that's awesome. I'm happy for you," I tell him.

"Thanks, man. It was early in the trip. I was near Boston and stayed at this Couchsurfing spot. This woman who owned the place had a friend, and so her and I, we hooked up," Jake then says, a sheepish smile pervading his face.

"You lucky bastard," I reply, failing to hide my envy. "I only got laid once on this trip so far. I thought it'd be a lot more frequent, but it just hasn't happened."

"Just wait until you get to New York," Jake then says.

"What's so special about New York?"

"You'll see. The women out there are different."

After the ninth inning expires and the fireworks are over it's time to go home, but Sam has other ideas.

"I'm going downtown," he proclaims with more force than the homerun back in the fifth inning.

"Bro, you're not going by yourself," I tell him, hoping he will stay in for the night.

"I might have to," he responds defiantly.

I know Sam will indeed go by himself, but he already has been drinking, and an early night sounds heavenly.

"You don't have to go," he assures. "I'll be fine."

"I'm not letting you go by yourself," I say, still hoping he'll eventually change his mind.

"I can drive you guys downtown," Craig offers as we drive home from the game, and it's then that I've signed up for a night out.

No part of me wants to go spend a bunch of money and get wasted, but if the roles were reversed and Sam was in my precarious position, he would do me the favor.

"I have to tell you something," the talkative and sagacious Craig says on the way downtown.

"What's that?" Sam asks.

"I'm not your father, I'm not your preacher, and I'm not your counselor. I'm not going to tell you how to conduct yourselves tonight." He pauses for a deep breath. "But goddamn it, do not bring a woman back to my house."

"Who do you think we are?" I chuckle.

"I'm just saying. You two are good guys, but I know how young men think."

His sentiment isn't shocking. Craig is one of those guys who always has a story. It doesn't matter if it's breakfast, lunch, or right as your eyes are about to close; he's going to tell you a story.

He never spares any details either, whether the subject is his failed marriage, the sour culmination of a fourteen-year relationship with a woman he biked across the country with, or a plethora of other narratives that quickly left me wishing life wasn't so complicated.

Yet the most noteworthy story all weekend has been the recitation of his thoughts as he sat in a hospital bed after a heart attack.

"I just asked God for forgiveness. It didn't matter how much money I had or anything like that. I just wanted to see my kids again. I just wanted to live, so I asked God to forgive me for the sins of my youth," Craig had said with so much force, I knew his life hadn't gone perfectly.

"Have a good night," I tell Craig as we exit his car.

"Remember what I said, fellas," he responds.

"Don't worry. We aren't bringing any women back."

Sam and I begin our night at a bar filled with people our age. During the first round, Sam starts talking with two dudes while myriad babes pass by. My patience is wearing thin because my balls are achy, and I start feeling insecure about both giving and receiving zero attention from the opposite sex.

In the middle of me slurping another Grey Goose Red Bull, a creepy guy loaded up on booze touches another dude's girlfriend, and then all hell breaks loose.

"You need to leave," an old bouncer says to the handsy jerk.

The creep looks at him, then back at the bartender without saying a word.

"You've had too much. It's time to go," the man behind the bar tells him, but still the creep does not budge.

"Time to go, now!" the bouncer says with much more force.

Again, the creep does not waver. Tired of the creep's petulance, the old man grabs his arm.

"Don't touch me," the creep demands.

The old man does not relent and instead grabs the creep's throat, inciting a skirmish. A horde of people rush over, but not even three bouncers can drag the creep out of the bar. They wrestle and smash into tables until finally five people are able to push the creep outside.

"What the fuck was that?" I whisper to myself.

I finish my drink and head outside, only to see Sam making conversation with the very same creep.

"Get the fuck over here," I yell at Sam, pushing him in the opposite direction of the douchebag.

"What's wrong?" Sam garbles.

"Long story, bro."

We head to a place down the street that's packed with women. It also has a clubroom with a DJ that puts most bars to shame.

"You want another drink?" a clearly intoxicated Sam asks after a few songs.

"Grey Goose Red Bull," I confidently beam, my eyes scanning the dance floor for an invitation.

Sam comes back with four Geese, so I have to chug the one I already have just to hold the other two.

This place affords many opportunities for the mind to wander. Scantily clad women of all races roam freely, but as time passes nothing other than a few intrigued glances come my way. No East Coast liberated women drop by to grab my ass or whisper over the deafening music into my ear. The narrative that out east is more liberated is not playing out in my favor. I become frustrated, not only at the lack of attention, but also at myself for doing absolutely *nothing* to improve my circumstances.

Sam is clearly having the best time. He dances like no one is watching, at the same time hitting on random black chicks who aren't interested in his advances. But he's putting me to shame because he's living his life, and I'm merely a passenger to mine.

And then Sam really starts ripping up the dance floor. A small crowd circles around him to watch him perform, but during one spin Sam loses control of his drink, fumbling it to the ground.

His head swivels to gauge the reactions of the onlookers, but it's not three seconds later before he breaks into a new routine, which elicits a throng of cheers.

More people push through to see what's happening, and I get pushed further away from the dance floor. I haven't lost hope of finding a woman, but still I do nothing, opting instead to apathetically stand in place like my legs are frozen.

"You're a pussy. You're a fucking pussy," I keep whispering, my body shaking with anger and frustration.

Sam returns a few songs later with more drinks. Tonight has become too much.

"You good?" I ask him.

He nods, but his eyes aren't symmetrical anymore.

Time passes me through a vortex into a dangerous state of irritation. I want sex and validation, but at this rate it will never happen, and now Sam needs my attention.

"This way," Sam garbles, motioning toward the outside patio where a live band is playing.

When an attractive woman brushes against my shoulder without acknowledging me, control is lost.

"We got to get the fuck out of here," I tell a discombobulated Sam, who by now can barely stand.

He nods his head in agreement, and we begin walking out when suddenly my right arm is being yanked out of its socket. I turn around to see Sam is grabbing me for balance.

"Jesus fucking Christ! Get your shit together!" I yell, but he doesn't hear.

We exit the venue and find a spot on the sidewalk to order a car.

"Do you want to get the Uber or should I?" he mumbles.

"I got it," I say, knowing he has spent too much money already.

I flip through the app until securing a ride, simultaneously perusing the bustling sidewalks for a straggling woman who might see me as her last-ditch effort for release.

I hate myself for being so pathetic, for spending the entire night not concerned with Sam or a good time, but instead forcing my gaze onto women. I have felt low many times, and this is easily one of my lowest.

The Uber arrives and brings us back to Craig's house. We quietly maneuver upstairs.

"Did you have fun tonight?" Sam sarcastically asks while lying on his belly on the air mattress below me.

I pretend to be asleep to avoid a 3:00 AM shouting match similar to the countless other ones we had back in Minnesota.

"I know you can hear me," he continues when I don't answer.

"Bro, I'm tired, but I did have fun," I deflect.

"Doubt it. I don't see any women here."

Sam laughs as he says this, and then burps up a little of that night's liquid. "I know you, Q. I know you."

His words ring in my head until my eyes mercifully shut.

CHAPTER 22

"I think I pissed the sheets," Sam says the next morning, his climb out of bed a laborious affair riddled with long groans.

"What do you mean you think you did?"

He holds up the sheet, and like you might see on a five-year-old's bedspread, a large circle of liquid is soaked onto the fabric.

"Oh, fuck me. We got to do something about that, dude, and I have to drink some water," I say, my stomach wrenching with liters of Red Bull still lodged inside.

After Sam quickly and covertly shoves urine-soaked sheets into the washer, we say goodbye to Craig and Sue.

"Thank you for everything," I tell them.

"Not a problem. Come back anytime," Craig says.

I begin wondering if I'll ever get laid again on this trip, yet altogether feeling much better now that the sun is up and thoughts of women can be delayed a few hours.

"Before we left, Craig told me he gave Jake a ride to the border," I tell Sam as we ease over one of the first few hills.

"That kid," Sam mutters, too hungover to voice his opinion.

Being hungover sucks, but Sam is on a level I have not seen since he called his old neighbor to propose they settle their differences by entering into a friends-with-benefits arrangement.

"Jesus Christ, kid. What the hell is the matter with you?" I scream when Sam and I nearly collide at a stoplight.

"Huh?" he mumbles.

"Sam, you have a fucking drinking problem."

"Sorry," is all he can softly mumble.

I no longer feel safe. Sam is going to end up in the obituaries if things don't change fast.

When he later misses a turn, my sense of fear is further heightened. "Do you know where we are going?" I ask.

I'm letting my emotions get the best of me. We only have twenty-five miles to ride, but at this rate we won't make it unscathed.

"I know you're not feeling well, Sam, but you have *got* to try to pull it together, man."

"I know, I know," he laments through heavy breathing.

The miles tick slowly away, and eventually Sam begins to take his place in the pecking order by speeding ahead. Two hours later I see the North East city limits sign. The person who came up with the name for this city must have been concussed because it's in the northwest corner of the state.

Five miles away from the destination, Sam pulls over, and we begin joking about last night. None of the bullshit matters, and it takes a healthy helping of Sam's convivial smile to realize as much.

"I think I drank too much," Sam says.

"You did, but you usually do," I jab. "But your dancing was on point. You had half the club cheering you on."

"I did?" Sam asks.

"Yeah. It was fun to watch."

There is still a lump of Red Bull lining the walls of my stomach. I want to poke my belly and let all the air out.

Right now, drinking has begun to feel like a chore. When my relationship ended, I began to see drinking as a way to get laid. And then I started to despise drinking because almost every time I did indulge, I never got laid. Sadly, this didn't mean I stopped. I still partook in alcohol, if for no other reason than I wasn't confident enough to tell people I no longer enjoyed getting drunk.

With drinking, I had a different indoctrination than most of my friends. It started one summer when I was hanging out with my cousin and her attractive but bitchy friend Katie.

We were in northern Minnesota, secluded in a small cabin town where people got way too drunk each time Independence Day rolled around. That night, being the fifteen-year-old idiot that I was,

THE LONG ROAD EAST

I assumed drinking alcohol would ingratiate me with the older and more mysterious Katie.

Since I had no idea what booze would do to me, I took a few small swigs from a blue Gatorade bottle. The cheap vodka tasted awful, but since I already tried some, I figured it was best to pretend that the vodka had me feeling right as rain.

"Feeling okay, Quent?" my cousin asked.

"I think so," I replied, assuming that since I hadn't face-planted in the ground, all was well.

"How much did you drink?" Katie asked.

"A decent amount," I told her, holding up the bottle for a visual reference.

They both laughed at me.

"There's only a few sips missing," my cousin said while roaring in laughter.

"But I feel something," I tried to assure them both.

"Sure you do," Katie mockingly smirked.

Intent on losing my virginity to the coldhearted Katie, I drank some more. It wasn't enough to get plastered, but enough to momentarily escape reality.

"Woah, easy there," Katie said when she saw how much I consumed.

She rushed over and touched my arm, so in my deluded brain I was halfway to a blow job.

I still don't know why I was into Katie. She was rude to my cousin and every other adult she interacted with. She rolled her eyes whenever I expressed an opinion, and she spent all day texting a dude named Chad who never replied fast enough to her texts.

"I know he's not doing anything. God, he's being so difficult," Katie often told my cousin, and I always laughed before wantonly trying to slide my way into her good graces.

Even though most people couldn't stand when Katie opened that massive orifice under her perfectly symmetrical nose, she was sexy, and that made up for all of the annoying aspects of her personality.

As the blue Gatorade mixed with cheap vodka penetrated my senses, I fantasized that Katie and I would go off into the poison ivy-laced woods, and she would turn me into a man. I'd then come out with rashes on my ankles and go back to school the next fall with the swag necessary to make my female contemporaries pine for my knowledge and experience.

The only problem to this fairy tale was that Katie wasn't interested. In fact, she hated me, likely because the first time we met I creepily took photos of her and then sent them to my friends and told them the woman popping up on their phones was my girlfriend.

Yep. Could have planned that one better.

Later that night, while my cousin talked about Lil Wayne's new album and Katie eagerly waited for her phone to ding, I began my more aggressive pursuit.

For most people, this might involve a slight touching of the thigh or a smooth transition into conversation. But for me, I deemed trying to fall into the firepit to be my best chance at getting laid.

I took one more lazy swig from the plastic bottle and then stumbled around the fire, the entire time extremely mindful of where my feet were positioned.

"You know, this stuff is pretty good," I told Katie.

"What the hell are you doing? Get away from the fire," Katie said, yanking me away from the fire.

"No, I like the fire," I tried to assure her.

"Quent, sit down. I don't want your parents to yell at me."

"They're cool. I've done this before," I lied.

Katie then grabbed a bottle of water. "Drink this," she said, shoving the bottle in my face.

"Maybe I will," I said, then chugged the entire bottle.

After this, my act began to fade, and then it morphed into desperation. Taking one last shot at glory, I decided it was best to pull my pants down.

When I did, my cousin turned away, and Katie reached for a camera.

"Oh my god," she yelped.

I turned my head thinking maybe there was a god looking out for me.

The flash boomed at me and then I heard another laugh, so I pulled my pants up.

"Wow," Katie began, turning to my cousin. "His dick is so…"

"Please tell me it's gone," my cousin said.

"Yes, it's gone," Katie laughed.

My cousin removed her hands from her eyes.

"His dick is so small," Katie then said, at which point both her and my cousin burst into a rollicking joy of laughter.

"You need to go to bed," my cousin said once she finally caught her breath.

That was the last time I was allowed to hang out with the older kids.

Sober cabbing sucks, especially when everyone around me is hammered and throwing their arms around my shoulders and telling me how much they love me.

Of course, someone has to be responsible, but one summer I found myself behind the wheel more often than I would have preferred. That year, I worked in a nondescript small town that hated outsiders. The weeks were long, and the job paid less than a low-rung fast-food joint, but finally, after ten weeks, it was time to leave.

My coworkers and I went out to celebrate, but once again I was in charge of staying levelheaded. I mercifully sipped on a glass of orange juice all evening until two of my coworkers agreed to call it a night and head back to the dorms with me.

One was named Annie. She was highly religious and thought that a woman having an orgasm was a sin. Ironically, that night during karaoke she chose to perform a song that talked about how if the lord didn't save her from her bodily urges, she was going to sacrifice her purity for a man with easy eyes and a nice smile.

The other was a big girl named Bertha. Not only was she notorious for wearing low-cut shirts in front of teenagers, she also was prone to hooking up with any guy she could get her hands on.

The combination of these two women in my car meant contention was inherent.

As the three of us were peeling out of a random bar's parking lot around midnight, Bertha started mumbling inaudibly.

"What did you say?" I asked.

More incoherent words spewed from Bertha's mouth.

"Did you hear what she said?" I turned and asked Annie.

"No," Annie said, her expression growing more concerned.

"Bertha, what did you just say?" I asked more adamantly.

A pause ensued, and Bertha writhed around in the backseat like she had a runny tummy.

"I was raped," she then said more clearly.

I pulled the car into a different parking lot, unsure how to react.

"Wait, what?"

"I was raped," Bertha said again.

"By who?"

"I don't know. Some guy."

Not understanding much at this point, I probed further. "Well, where did this happen? At the bar?"

"No," she whispered.

"Well, where? I'm confused. You have to help me out here."

Annie was looking at me in disbelief, the beginnings of Bertha's story not making sense.

"Bertha, I'm going to call the cops. I really don't know what to say or do," I told her.

"No, don't," she said.

"Why not?" I questioned.

Bertha didn't immediately answer, then proceeded to explain that some guy dragged her out of the bar and down to a massive lake next to a hotel, where he then committed the crime.

Trying to do the right thing, I called the police, who quickly arrived and said they would be taking Bertha to the hospital.

"You can come to the hospital," the officer informed me.

"How far away is it?" I asked.

"About forty-five minutes."

I looked to Annie for guidance, but even though it was almost 1:30 AM, I knew we couldn't just drive home and leave Bertha. Annie and I then drove to the hospital in relative silence.

After arriving, we were told to sit down in the waiting room. As soon as I tilted my head back, Annie became unhinged.

"Bertha did this to herself," she deadpanned, and I thought Annie might take out a Bible before beginning her next verse.

"I mean, I don't know. That's kind of a rough thing to say," I told her, swiveling my head and hoping Bertha would soon come out and save me from this sermon.

"Her behavior…," Annie continued, launching into a vitriolic diatribe that painted Bertha as a disgrace to the heavenly father.

"Maybe we should just wait to see what happens," I told Annie after she ran out of breath and malice.

Soon I got a call from one of my friends who was still at the bar.

"There's some guy out here crying," he told me.

"What do you mean?" I asked.

"He said he had sex with a girl, and then after she told him he raped her."

"Jesus fucking Christ," I replied. "Did the guy say who the girl was?"

"He didn't know her name. Just that she had blonde hair and was overweight."

The puzzle was slowly coming together. I could only assume this man's story wasn't a coincidence.

More time passed, but my eyes could not close. Annie had slowed her condemnation of Bertha, but as the early morning turned closer to actual morning, I just wanted to figure this situation out and go home.

With my phone reading 4:12 AM, Bertha walked out as a nurse followed behind her with a bag I assumed was a rape kit.

"We have to go to the police station," Bertha told me, and I rolled my eyes because if I didn't they might just close shut for the evening.

Nearly an hour later, we got back to the police station. Annie and I sat silently in an empty room while Bertha and a police officer went into a room next door. Within a few minutes, shouting started coming from the other room. It became evident something was not right here. Annie looked at me, and without speaking we both assumed Bertha had duped us.

A few minutes later the officer came out of the office as Bertha lurked behind him with a blank expression on her face.

"You guys can go home now," the officer said.

No words were spoken on the ride home that night. Bertha sat quietly in the back, and I took note of the morning sun that was slowly rising.

After a few hours of sleep, I woke up exhausted. When I hopped into my car that morning to drive away, there was not an ounce of shame on Bertha's face. She looked as chipper as ever, like the previous night was everything she could have hoped it might be.

"Are you okay?" I hesitantly asked her.

"Of course. Why wouldn't I be?" she responded.

"I just figured with everything that happened."

"Oh, you mean last night? Yeah, that was all just a misunderstanding."

Bertha's words were so cold and empty, almost as if there were no feelings attached to the sounds coming from her mouth.

"I better get going then," I told Bertha.

When I got in the car, my friend shook his head.

"Maybe we should stop drinking?" he asked, by then fully aware of how my night played out.

"Actually"—I said, then putting the car in gear—"that might not be the worst idea in the world."

***

That evening in North East, Sam and I meet a man named Dan. His house is at the far end of a neighborhood, one empty lot separating and pushing his property to the outskirts of the development. A Hummer sits in his driveway, and when he shows us his garage full of toys, it's apparent this guy has poured a lot of money into *things*. We walk inside as Dan shows us the rest of his home.

"Do you want some grape juice?" he offers when we sit down in the living room.

"That sounds good. Thank you," I tell him.

Dan then takes out a small can of Welch's grape juice from his fridge and hands it to me.

"These cans are kind of unique," I observe.

"Well, I work at Welch's."

"Really? How far away is that?"

"Maybe a few miles that way." He points as if I am going to follow the direction of his index finger, then explains that North East is the headquarters for the grape juice conglomerate.

"Cool," I note, and then his cat brushes against my leg.

"Sorry about her. We don't get a ton of visitors," Dan explains.

"No worries. I just hate cats," I say.

Dan blinks a few times. "I don't know what to say to that."

"You don't have to say anything. I just wanted to mention that for some reason."

Not long after, we are debating dinner options when Sam suddenly ejects a massive fart.

"Whoa, who's that asshole behind me?" he cackles, looking behind him even though he is pressed up against a window.

Neither Dan nor I laugh, so Sam's head sinks to the ground.

"Sorry, sir," he mumbles.

Sam's shamed comment drives us to burst into laughter. "You're nasty, bro. Save that for the shower," I remark.

There are a few other people at the small bar we visit. Most of them are sipping on Yuengling tap beer and looking like they don't know where they went wrong in life. The waitress comes over and takes our order.

Dan and Sam drink the pale ale while I munch on a burger, and over time Sam continues to order more beers. He's either running away from the residual pain of last night or sprinting toward the hell that will come tomorrow.

Once Sam empties a fourth glass, we pay our tab and leave. I'm not looking forward to going to bed tonight because I'll be splayed out on the floor and susceptible to the cat's advances during the middle of the night.

For the next two hours my skin rubs against the harsh carpet. Today wasn't very fun. We have to get to New York.

CHAPTER 23

"Fill your tires, Q," Sam instructs as we gather in the driveway the next morning.

Dismissive of Sam, I refuse to pump air into my tubes because I don't want to cave to his condescension. But once my defensiveness simmers, I'm down on one knee following orders because to do otherwise would be foolish and might lead to a flat tire.

I begin fiddling with the valve on my tire. The connection nozzle from the pump won't attach to my valve.

Cue the headache.

"Got it there, chief?" Sam chuckles, looking at Dan to see if he derives the same amount of joy from my struggle.

"Just give me a second, Sam," I snap.

"Do you need some help?" Dan asks.

"That would actually be awesome. I've never used a pump like this."

"It's a pretty basic tire pump," Dan assures.

"Maybe I'm just an idiot then," I chuckle, trying to deflect my ineptitude off as ambivalence.

Dan kindly fills my tires with air. We shake hands and then mount our bikes. "See you later, Dan," I say. "Thank you for everything."

As Sam and I begin, the Monday morning chill acts as the cold shower I have forever refused to take. We are now leaving the flat Midwest and heading into the mountains, and soon we approach a sign that says we're in New York.

Sam stops, and our cameras begin flashing away. Not much has changed except the vanity of being in New York suddenly feels important.

THE LONG ROAD EAST

"I cried when I saw that photo," my mom told me months later. "I was just so proud of you guys."

Reaching this sign feels like an accomplishment. Part of this has to do with the fact that we just rode our bikes all the way from Minnesota to New York, and part of it is that sometimes I can't help but be taken aback by the way the world works.

I knew Sam and I would reach this point when we set out. It would have taken death or an insurmountable financial crisis to prevent this moment from happening. But it still feels sweet, that fleeting rush of energy that occurs when something significant takes place. For as nice as this feeling is, the hope is that there are more to come.

Once we get over the moment, we hop back on our bikes. Some idiot on a motorcycle is plopped directly on the shoulder only one hundred yards ahead. *Get the hell out of my way, you fuck,* I want to say to the middle-aged man with a ponytail. If he doesn't move, I will have to swerve into the right lane of a county highway. Doing this won't be the end of the world unless a car doesn't see me and proceeds to crash and shatter my tibia.

To this point, we have moved through Minnesota, Wisconsin, Michigan, Ohio, and Pennsylvania with relative ease. There were tough hills in Wisconsin and torrential rain in Michigan, but we lucked out in Ohio by staying in the northern part of the state and avoiding all the hills down south. And we didn't spend enough time in Pennsylvania to fully grasp what type of terrain it had to offer. I had no reason to think New York would drastically differ from the previous states.

We stop once again after twenty miles.

"Question," Sam says, looking at me and then back at his phone a few times. "We are staying in Buffalo tomorrow night, correct?"

"Yes."

"But we are staying in Springville tonight?"

"That's correct."

Sam pauses once again. I know he has an issue, and it most likely derives from the sign a few miles back that says Buffalo and Springville are both thirty miles away.

"You didn't happen to see that sign back there?" he asks.

"Which sign?" I ask as if there was no sign back there.

"Look. We are going out of the way to get to Springville, and Buffalo is the same distance. You see what I'm saying?" he asks, pointing at a map on his phone.

I see what he is saying, though I don't let on as much.

"We are essentially adding more miles for no reason," Sam complains.

"That's the way the hosts played out," I say.

"How about we go to Buffalo tonight? I'll pay for a hotel room."

"You don't have to do that," I retort, wanting to stick to my previously devised itinerary.

"Your way doesn't make a whole lot of sense though," Sam says with more conviction.

We go back and forth on logistics, but Sam's right—we don't need to go out of the way to eventually arrive in Buffalo the next day. My route adds another day to this trip. That, to me, is the beauty in all of this.

Springville was always on the docket. A few days prior, Sam and I changed course and collectively agreed we were going to see Niagara Falls, and then hike horizontally across the state to New York City. What we didn't think about was how far it would be from Niagara Falls to the big city.

I had also earlier labeled Springville as a stopping point. Had we never decided to go to Niagara Falls, Springville would still have been on the itinerary. This was because I set up a full course before we left. That route was going to get us there in four weeks, provided we rode every day and hit all the benchmarks. But I figured interests and desires would change as we settled in, so I never got married to the planned route.

What Sam didn't understand is that when we made variations, this put me in scramble mode to find Warmshowers hosts. And finding hosts wasn't easy. It wasn't rocket science, but I didn't just press a button on my phone and get linked to an amicable human being. I did my homework on people, looking at what they offered in terms of housing, food, nearby stores, etc. This took time, and so whenever

THE LONG ROAD EAST

Sam lectures me on the pragmatics of the route, I feel compelled to push back.

To be fair, Sam usually throws me nice compliments when we describe to hosts our process in getting from city to city. As he eloquently says, "If Q was in charge of getting us there, we would never make it, and if I was in charge of finding us spots to sleep, we would be sleeping on the side of the road most nights."

Not wanting to argue any more, Sam relents, and we continue to Springville even though it's out of the way. When we later pause at a stop sign, a man on a lawn mower takes a break from mowing to have a quick word.

"Where are you from?" he asks at one point in a conversation that's running too long.

"Minneapolis," I say.

"Oh, Minneapolis! I drive through all the…"

He goes on a rant that Sam eats up because this guy also has a beer belly and owns a John Deere. The sun is fast setting, and this random man is taking up too much time for my liking. I give Sam a look of contempt.

"This has been great, sir, but we should get going," Sam tells the man, and then we're off.

Later I come to an intersection. One sign says Springville is eight miles to the left. I remember Sam saying stay straight but then doubt my own memory. I look at my phone, and of course have no signal.

"Goddamn, T-Mobile," I curse.

Sam never makes a turn without acknowledging I know to follow, so one could assume staying straight would be the best course of action. At the same time, I don't want to doubt the green sign before my eyes.

I follow my gut and the sign to Springville, but a few miles down the turn there is still no sign of Sam. I pass a family saying goodbye like it's the last time they will ever see each other. The sun is becoming less and less apparent. I'm now apprehensive about my decision, but when I check my phone a few bars fortunately pop up. Time to call Sam.

"Hello!" an angry voice answers.

"Did you follow the sign to Springville?" I ask.

"Huh?"

"I followed the sign to Springville. Am I going the right way?"

"Yes!" the spastic bird hollers, then hangs up.

"Hmm, that guy is a little angry," I say to no one.

The reason for Sam's tone is concerning, but then I see a range of mountains, and it all makes sense why he was in no mood to talk. The road becomes steeper at every curve, reverting all the way back to that stretch in Wisconsin that made my lumbar howl for mercy. There are only a couple miles to go, but they are as grueling as the Canadian wind I endured back in the winter of 2015.

"You called me when I was going up a hill," Sam says when I find him sucking down a green tea at a gas station.

I try to hold in my laughter until my chest hurts. It's difficult swallowing a bellow.

"Yeah, those hills were not fun," I snip.

I still don't understand what Sam thinks we are missing in Buffalo. He hates the hills, but they aren't going to kill us, and he's riding excellently.

Outside an apartment complex, we meet Forster, our Warmshowers host. He is a man's man, his graying beard punctuating a face most divorced women would drool over. There is an aura surrounding Forster, one that suggests he'll be your friend, but try anything weird and he'll kill you.

"I'm not a great cook, so I apologize, fellas," he says while we eat rice and beans sprinkled with Frank's hot sauce.

Cut from the same cloth, his words are endearing.

"I hear you're a writer," Forster proceeds to mention.

"I am," I tell him.

"Well, what's your book about?"

"In so many ways, man, it's about my time in college, and how biking transformed me."

"Is there sex in it?"

I laugh. "Yes, there is sex in it."

THE LONG ROAD EAST

"I just finished this book," he transitions. "This guy writes about sex also. He talks about all the women he banged. One of the women he sleeps with, their shared goal was to have sex in every position imaginable."

"That sounds wild."

"I shit you not. He talks about shoving his dick into this woman's armpit."

I stop to make sure food doesn't explode from my mouth, unsure if having armpit sex warrants laughter or disgust. But that's how our night begins, and it evolves from there.

"I have to tell you boys about Uncle Bobby," Forster says.

He jumps into a layered discourse on Uncle Bobby, the passion in his words and eyes as intense as the fateful day Uncle Bobby showed up at his doorstep.

"So this fucking Uncle Bobby. And his son, ah, I don't remember his name. They were my first Warmshowers guests. Dudes could burn. Uncle Bobby was 750,000 miles into his goal of one million for his lifetime. His son was a sponsored rider. Anyway, they had just gotten off a long day. We're talking 125 miles."

"Damn," Sam notes.

"Right?" Forster continues. "Nice guys, but Uncle Bobby was here for maybe a minute, and then he gets buck naked in my living room and walks straight to the shower. I'm looking at him in disbelief."

The story doesn't seem that egregious until, "And these guys must not have eaten all day, because they got here and literally ate *everything*! They ate so much food that we went to go get more, and they ate all of that! And then the next morning, I had gone to work, and they took all my stuff for PBJs."

Sam and I stare at Forster, bewildered and now conscious not to eat his entire pantry. Later that night, more banter ensues.

"I was in Alaska for a year, and I didn't get laid the entire time," Forster reminisces. "All the women there were used to dealing with douchebag military guys, so if you weren't trying to get married, they wanted nothing to do with you."

"I think I'd die," I joke.

"That was the most masturbating I ever did in my life," Forster reveals.

More stories on the escapades of Forster's primitive years are revealed as the night evolves.

"My buddy and I stopped in a town one night, and we are at this ATM. He starts hitting it off with these girls," Forster begins. "Long story short, he is going home with the hot one, the other hot one is going home alone, and so here I am with the fat one."

"Lucky you," I pipe in.

"Right?" he says, his eyes rolling all the way back to that night. "And so my buddy is definitely leaving. I'm kind of drunk, but not really drunk, and I don't want to sleep in the car."

"*You can come back to my place*," the rotund woman in the story offers.

"Oh, man. So here I am, not into this woman at all. We are at her place, and she's looking at me like she wants me to make a move, but I just keep telling myself I can't. Then she comes closer and touches me. *Oh fuck*, I think to myself."

Forster stops to adjust his glasses.

"Now I'm looking at the table, and there is a bottle of tequila, and I just go over there and start chugging out of the bottle. There was no way I was banging this fat chick anything less blackout drunk."

"At least you're honest," Sam tells him.

"Next morning, I wake up, hungover, and she wants round two. Finally, I say no. Make up a bunch of excuses about not feeling well and thankfully got the hell out of there," Forster finishes.

"Were you really too hungover to go again?" I ask.

"No, but, dude, I mean, this chick has to know the score. I chugged tequila in front of her face before we had sex."

"Jesus. I thought it was only guys that were that desperate," I say.

There is a charisma to the way Forster speaks. As the night grows darker, so do his stories. He goes on to explain his divorce and how ruthless his ex-wife was.

"It just didn't work out?" I ask.

"It's hard to stay married to a woman like her."

THE LONG ROAD EAST

"What'd she do?"

"A lot of things, but the one that would bother me the most is during our divorce she would sneak this guy over to my house, the house I paid for, and then lie to me about doing it. One day I finally saw him on this Strava app, because I knew he was a biker. The dude went to my house, spent twenty minutes there before she had to pick up the kids, and then left. Can't imagine what they did in those twenty minutes, right?"

The hurt in Forster's eyes is evident. As he speaks, his eye contact never wavers. I can see the pain he felt when his rotten ex-wife was ruining their marriage.

"All I can say is that women are brutal. If you piss them off, they will try to ruin your life," Forster explains.

"Jesus Christ. Fuck getting married," I say as the night closes. "That shit sounds brutal."

Forster looks into both our eyes before standing up to go to bed. "It's rough. Hell hath no fury like a woman scorned."

CHAPTER 24

"Don't say it," Sam hisses as we trudge through the morning.
"Don't say what?"
"Come on, you know what I mean."
"I have no idea what you're talking about, Sam."
"You know."
"No, I don't know. Please enlighten me."
"Because we had a good time last night."
His words still a mystery, I seek further clarification.
"You don't remember me not wanting to come out here?" Sam asks.
"Ohhhh!" I exclaim, throwing my head back. "You mean because we didn't go to Buffalo, and Forster ended up being a good dude. Bro, don't even worry about it."
"I just don't want you giving me shit for this later, Q."
"I won't. All's well that ends well."
As much as I like being right, we've been on the road too long for pettiness to develop from something as minuscule as a route disagreement.
"I'll tell you one thing though, bro. I would not want to get married after hearing what I heard last night," I say. "His ex is a fucking bitch."
"Yeah, but she's just one woman. Don't confuse women like her with all women," Sam says.
Getting to Buffalo is a breeze. We eventually pull up to a guy named Chuck's house. He seems normal, a painter by day and valet by night. He has dark-rimmed glasses and a collection of bikes in his entryway.
After he shows us around his place, the three of us agree it is time for lunch.

THE LONG ROAD EAST

"We have to bike though. I don't have a car," Chuck says.

"Perhaps we could walk," I offer.

"It's better if we bike. You can take all your gear off though," he mentions.

We soon are riding through a cemetery into oncoming traffic, stopping in a roundabout before a few miles later locking our bikes onto a stop sign just outside a restaurant.

"Refills cost extra," the waitress informs Sam as he slugs down his third Sprite of the meal.

"I suppose I'll switch to water," he tells her.

Back at the house, Chuck talks so much about the inane "sport" of bike hockey that I nearly fall asleep.

"You ever played?" he asks.

"I don't even know what it is," I tell him.

"There is a YouTube video of it," he says, then goes on YouTube and begins subjecting my eyes to thirty minutes of the most excruciatingly boring activity I have ever witnessed.

"How about dinner?" I interject halfway through the second video.

We again jump on our bikes and ride to another part of town, settling on a local burrito spot that pales in comparison to my beloved Chipotle.

"What do you guys want to do tonight?" Chuck asks as we carb load.

"It's your city, bro. We're just along for the ride," I respond.

"We could go to this arcade. They have pinball. And a bar."

It's not July fourth, but Sam's eyes crackle at the mention of a bar. When I hear pinball, I want to check Chuck's ID to make sure he isn't twelve.

"Will there be women at this place?" I ask.

"I don't know. I have a girlfriend, so I don't really know which spots are good for that."

I sigh because whenever a guy uses that as an excuse for not knowing fun places, it's just a mask to cover up the fact that even if he were single, cool spots would still elude him.

"I am down for pinball," Sam says, even though I know he likes pinball as much as Ellen DeGeneres likes men.

"Chuck, I'll be honest, I don't really like pinball or arcade games," I offer, the idea of watching those two throw balls up a gutter or seeing children fight over a Pac-Man machine sounding awful.

From my vantage point, our time is better spent looking for single women.

After some deliberation, we agree to conserve money for the trip to Niagara Falls the next day and go home. I check Tinder, and there is a message from a chick who is way too put together for temporary impurity, but I still have to shoot my shot.

"Hey, Chuck. What would you say if I wanted to hook up with some Tinder chick tonight?" I ask.

"JESUS CHRIST, KID!" yells Sam as we turn up a neighborhood.

"What?" I ask.

Sam's inner priest is making a resurrection. "You won't go play pinball, and now you want to bring some random woman into Chuck's house."

Chuck is finding this whole sequence highly comical.

"I'm just being respectful," I tell Sam. "Besides, you just want to go get your beak wet," I counter.

"No. I don't need to drink tonight," he says in a tone only four years of friendship can tell is a lie.

We spar the rest of the ride home, and once my Tinder prowess fails for the forty-third time this month, I realize not much has changed since we switched to the eastern time zone.

"No luck, huh?" Chuck pipes in as another look of defeat forms on my face.

"She said she doesn't do those types of things," I whimper.

"Not surprising."

"The last time I was on the East Coast things were much different," I complain. "I don't get it. I thought women out here didn't want to get married."

Chuck shakes his head and laughs, clearly content with his lot in life.

"All women want to get married," he says.

CHAPTER 25

My balls are bluer than the Atlantic Ocean. To counter this, I am going to sit on them all day riding to Niagara Falls, one of seven things the world wonders about.

"Morning, Chuck," I say in between bites of organic cereal.

"Don't eat all of that, please," Chuck says.

"I wasn't planning on it, bro."

"I know, but that stuff is expensive. Maybe I should have told you to eat something else."

Perhaps I haven't given Chuck a fair shake. He's a good guy, but he is also an *amper*, meaning everything he does is self-described as the coolest thing ever. Whether it's talking about his job, house, or the fancy art school he attended five years ago, all Chuck wants to do is tell me how cool his life is.

I want to tell him I don't care, that if he was as awesome as he thinks he is, I would know, and he wouldn't have to cite a plethora of reasons to convince me of his legitimacy.

The three of us leave after breakfast and begin riding toward Niagara Falls. "Is your bike riding okay?" Chuck asks after I've fallen behind him and Sam.

"Yeah. My bike's just old," I say defensively.

Chuck gives my bike the up-down, then politely smiles.

"Do you have something to say?" I ask.

"No. Your bike is old though."

We get off a trail and enter one of the largest bridges in Buffalo. As we ascend a couple stories, I can see from the guardrail that it's a long way down. The rail isn't holding much in, and it will only take a tiny dosage of bad luck for one of us to go toppling over the rail.

Sam falls behind on the ascent because his bike weighs nearly three times as much as ours. On the other side of the top of the bridge, a spastic older man with no home has a conniption.

"You're supposed to use the other side of the bridge," he screams in my left ear as we shimmy past each other.

"Why was that guy so pissed off?" I ask Chuck at the bottom of the other side.

"Fuck him. The other side is even worse than this side," Chuck says.

Sam then pulls up.

"I see you're falling behind, bro," I crack.

"That's a long way up," Sam gasps, his breathing labored.

"Don't worry. I'm sure I won't beat you again," I say, but what I really want to tell him is that "I'm a better athlete, you sorry son of a bitch."

"Just like NHL," Sam responds, in reference to the countless hours I spent destroying him in a video game modeled after the preeminent hockey league in America.

As the three of us get closer to the mouth of Niagara Falls, the river begins running faster. Swaths of bugs shaped like giant grasshoppers attach themselves to every article of our clothing.

"What is with all the bugs, Chuck?" I ask.

"It's that time of season."

These bugs are irritating, landing on my chest, ankles, face, and a few times smacking against my lips. I brush a few off, but then dozens more take their place.

When we arrive at Niagara Falls, it is surrounded by tourists that, like Sam, take photos of everything. I walk down the stairs to get a better glimpse of what's below. It is so loud, and there is so much water running into the falls' basin that my head starts to hurt. I lean over the rail, the bottom miles away.

"You know people have went off into the mouth," Chuck shares.

"That sounds like a death sentence," I laugh.

"Sometimes it is. A few have lived. Others have died."

"What do they ride down in?" I ask.

"Wooden barrels."

"That's it?"

"Yeah. It's wild. They squeeze into these barrels and then ride down the river. Eventually they go over and into the bottom."

"I see. They die from the fall then?" I ask.

"Exactly. It depends where they land, and how they land."

"Wow," I say, looking over the guardrail once more to gauge just how far a fall would be.

Niagara Falls is cool, but after a few minutes it turns into everything else a simple Google search would have revealed.

"I'm not going to hop on a plane to come see this again," Sam mentions.

After lunch, Chuck asks us if we want to take a different route back to his home. "There are less bugs this way," he says.

"Is the route longer?" I ask.

"No. Pretty much the same," he assures.

A few minutes into this alternate route, and already I'm pissed. The bugs are worse. I start to fall behind, enraged by the perpetually disturbing insects that have attached themselves to my face and legs.

"Are you sure this way is better?" I ask when we stop for a snack at Jimmy John's. "It felt like it was all uphill."

"It might have been a little longer," Chuck cautiously reveals.

"How much longer?"

"Uh, like eight miles."

"Eight miles!" I holler.

Chuck has fallen off the deep end. These extra miles are not doing Sam and I any favors for tomorrow, but there's nothing we can do except get back to Chuck's place before the afternoon turns to evening.

"Where are we staying tonight?" Sam asks me as we make a left turn and close in on our destination.

"With a woman named Anne," I tell him.

"May I ask why?"

Sam's tone is annoying because he already knows the answer. "What do you mean why? Because we needed a place to stay, Sam," I respond with an equally distasteful tone.

Sam wants to stay with Chuck another night because while I am fuming over Chuck's antics, Sam is enamored with him, his bikes, and his ability to go to work every day and paint white lines on tennis courts.

"Chuck told me we could stay with him another day," Sam mentions.

"That's great, but I already have lined us up another host, bro."

Without responding, Sam pedals ahead.

Thirty minutes later, we finally get back to Chuck's house, and after an hour of wasting time pretending we are staying another night, I start to make explicit hints that it's time to go.

"Anne is waiting for us," I tell Sam while standing close to the door.

Sam must have eaten too much for lunch because just getting off the couch is a lengthy process for him, and Chuck can't fathom why I feel compelled to move on. He begins inching closer to me as I tighten Joad to the back of my bike.

"I've cancelled on people all the time when I've been on trips. It's just the way it goes," Chuck insists.

"That's not how I do things, Chuck," I say, looking fiercely into his eyes to cement my point.

"Hmm, okay, but do you guys need help getting to the next place?"

"We have a GPS, so it's not necessary, man."

"You can come if you want," Sam tells him.

Chuck grabs his bike and joins us for the two-mile jaunt to a different neighborhood in Buffalo. When we arrive, our new host Anne comes outside to greet us.

"I thought there would only be two of you," she politely questions.

"There are. This is our friend Chuck. We stayed with him last night. He'll be leaving soon," I explain.

"Hi, I'm Chuck," he says to Anne. I look back with frustration. Chuck should be pedaling away right now.

"It's been fun, Chuck," I say.

THE LONG ROAD EAST

With the facial expression of a child who just lost his dog, Chuck looks away dejectedly, then slowly begins pedaling up the road.

"Anyway, it's great to meet you, Anne," I say, turning back to her and walking closer to the house to shake her hand.

Sam and I break down our stuff and cram it inside the upstairs of her small house. I immediately sense Sam is dissatisfied we gave up a comfy night at Chuck's place to sleep in this cramped Buffalo rental.

It's hard to blame him. I'll be sleeping on a couch that half my leg hangs over, while Sam is relegated to the wooden floor. My planning looks suspect, but explaining to Sammy that the quality of Warmshowers is mostly random would be futile.

As we unpack our things, Anne comes out from the kitchen to offer us pea soup.

"I'm more of a meat eater, but thank you," I respond.

"There is a grocery store just up the road," she then counters, and soon I make the lonely walk to fetch some bread, a container of Oscar Mayer bologna, and chocolate chip cookies.

When I return, Anne and her friend are sitting cross-legged on the steps. Their expressions are inviting, but their conversation centering around their relationships contains less intrigue than biodegradable lab reports.

I try to be polite and hang out for a few minutes before ultimately feeling emotionally drained and heading upstairs. I'd rather deal with constant rejection on Tinder than pretend to be cordial with a random Warmshowers host.

CHAPTER 26

The next morning, I unfasten Joad from the rear of the house, where she safely spent the previous night strapped to a water meter. A gloomy day of weather awaits, but what's surprising is that upstate New York proves to be as barren as Michigan. As a clueless young man, I imagined all of New York to be glamorous, tall buildings and bustling cities the norm. But after Buffalo, it's back to barren wasteland.

The day goes quick, and soon it's almost over. The sky is perpetually overcast but offers no rain. Captivating dark trees litter both sides of the road, providing a calmness I haven't felt since last year. This isn't the New York I've always seen through television screens.

A few miles away from the destination, Sam veers around a bulky tractor moving too quickly for a safe passing. I follow his course, veering into the oncoming lane. Fortunately, no cars from the opposite direction come, allowing me to safely pass the old man inside who waves calmly as I ease around him and back onto the shoulder.

After arriving at our host Paula's address, three dogs of various sizes howl behind a large window overlooking the property. Two are big, and the third looks like breakfast for a bald eagle.

"Are we at the right place?" I ask Sam.

"The GPS says we are," he says.

I walk to the front door and peer in through a skinny window that's parallel with the door. The house is empty, and then the dogs come rushing to the window, their paws slamming into the glass and causing a noticeable reverberation.

Time to whip out my phone.

"Paula, it's Quentin Super. We are at your place now," I tell the Warmshowers host, kicking the stray pebbles in her driveway around in frustration.

"Hey, Quentin. I was tied up at work, but I should be home soon."

"No worries. I didn't know if we were at the right place."

"You probably are."

"You have three dogs, right?" I ask.

"Yep, that's them. Jasper, Junior, and Joey. They're harmless."

"I love dogs, so that's not a problem."

Then Paula shifts the tide. "We also have a cat."

"See you soon, Paula," I respond, not willing to reveal my negative feelings toward felines just yet.

Twenty minutes later, Paula leads us through her cluttered, unfinished back room that has random items sporadically placed throughout.

"Are you planning on turning this room into something?" I ask.

"My boyfriend and I are planning on it. Please, make yourselves at home," she says. "I know there isn't a ton of space, but I want you to feel comfortable."

"Thank you," Sam says before heading to the shower.

I lay down on my bed in attempt to process everything. It has been a long day readjusting to country roads. Consistency feels mandatory. If we have to spend more time in cities, that's fine, so long as that's the norm. The same goes for rural areas, but oscillating between both is becoming draining.

Instability has now become bothersome. Being on the road isn't as romantic as it was back when this trip began, a time when anything seemed possible. The passing weeks have begun to wear me down, not in a physical sense, but emotionally all the rejection, poor interactions, and couch hopping has made me numb to the difficulties of meeting new people. Now that process is transactional and, often, unenjoyable.

Sam comes out of the shower and sets his clothes on the bed by the window, the same one where the cat sleeps. He will be covered in hair the next morning, but right now his only concern is eating dinner.

"You ready to eat?" he asks before moving past the sliding doors and walking into the dining room.

"Bro, I'm always down to eat."

Paula and her partner Sigmund have prepared a fantastic spread. Meat, meat, and more meat, plus a dessert sit idly on the round wooden table. I fall into a food coma shortly after the meal, Paula's words becoming more difficult to process as the pound of meat sitting in my stomach begins to digest.

"You look tired," she says.

"I am. It's been a long day, Paula," I tell her, and with that head for the comforts of my bed.

My stomach sated and full of sustainable energy, the white comforter on the bed keeps me warm, subtly massaging my grimy, infection-plagued big toes every time they rub against the fabric.

CHAPTER 27

"I made pancakes," Paula says as I walk into the kitchen.

"Oh nice. What kind?"

"Apple. They're the best."

"Perfect. And how's your morning go so far? Ready to start your day?"

"Absolutely," Paula says.

I eat ten scrumptious pancakes as Paula explains that she can treat her type 2 diabetes better than her doctor.

"How is that possible?" I ask.

"I know my body better than he does," she replies confidently.

"Interesting. I don't know much about medicine, but I am surprised that you know more than your doctor," I remark.

"You wouldn't believe it by looking at me, but I'm actually healthy," Paula claims, her hulking presence casting a shadow over the counter she stands behind.

"Right. Well, look, Paula, I have to go to the bathroom. Thank you so much for breakfast. It was fantastic."

Sam is picking the last remnants of cat hair off his socks when I begin trying to usher him out the door. I really don't want to get stuck behind his morning routine that consists of a thirty-minute bathroom break and small talk with the host.

"This is disgusting," he says, holding up a black sock that looks like it was drowned in a litter box.

"Geez, yeah it is, but will you be ready to go soon? I have had enough of this place."

"Fuck me, this is gross," he says. Sam looks at the floor and then the dirty sock. "Yeah, I'll be ready. Just give me a minute. Gotta clean up this mess."

A short while later we begin our long trek to the city of Lyons.

"What did you think of Paula's claim about being healthy?" I ask.

"Look, I know where you are going with this, so don't even start," Sam says.

"What do you mean, bro?"

"Q, sometimes you just have to let people have their moment."

"Yeah, but, man, the things she was saying—"

"Look, obviously, Paula is a little ahead of herself, but again, there is no need to discuss this. I'm not going to fight with you."

"We're not fighting. We're just talking," I reply with joy, looking up from the road to see if perhaps we will end up bickering about someone else's life, but Sam begins riding ahead.

That day, as we ride out of Rochester, the GPS starts faltering. "This thing is on crack today," Sam complains as we bike in circles around a park.

A mile later we come to a construction sign. It's telling us to turn around. Another decision has to be made. We normally disregard construction signs, but this one is blocking a bike path.

"What do you think?" I ask in the tone reserved for questions with obvious answers.

"Let's just see what's ahead," Sam ushers, but he seems less sure than he historically does about these kinds of things.

A couple minutes past the sign, Sam pulls off and talks with two people seated underneath a bridge. The entire scene feels sketchy.

"I just talked to some people," Sam starts when he comes back, oblivious to the fact I saw everything he just did. "They said we should be fine."

"They? You mean the people hanging out under that bridge?"

"Yeah. Why? Is there a problem?"

"Do you actually trust *them*?" I ask.

"What do you mean?"

"What do I mean? Sam, you just asked two people underneath a bridge for directions, or does that not seem strange to you?"

"It'll be fine. Relax," he says, but Sam's intention to inspire confidence is drawing little faith.

We pull ahead a few hundred yards and are instantly met with heavy machinery and no way through.

"Looks like we are taking the detour," I acknowledge.

We turn around, and the two individuals have moved from the darkness. Both have lines on their faces, like the Javier Bardem character after he takes out his teeth in the James Bond movie *Skyfall*. Their faces look ragged, to the point I grip my handlebars tighter and adrenaline kicks in. They stand next to the trail to watch us pass, their eyes revealing a desire for something greater than what the present has offered them.

"We're lucky they didn't try to stab us or something," Sam says after we get back to a road.

"Sam, you're too trusting. We're in goddamn New York."

"So what?"

"Bro, you can't be asking people who live under bridges for help. It's that simple."

"They didn't do anything though, and you don't know where they live."

"Sam, you saw how they looked. I shouldn't need to say anything else."

Out here, my feelings of vulnerability have amplified. I don't have a logical reason for these emotions, other than every day we move closer to New York City it's conjunctively exciting and worrisome. I don't know what to expect, and that notion doesn't sound as romantic as it once did.

"What would you say if I met up with some chick for a couple hours?" I ask Sam during a break.

"You know how I feel about hypotheticals. Talk to me when you have actual plans."

"But I don't want to make plans and leave you out of the loop," I urge.

"I don't want to talk hypotheticals, Q."

"So you'd be cool if I asked you to wait for a few hours while I did my thing?"

"Well, no. I'd like a heads-up."

"But that's what I'm doing now!" I holler, completely confused.

Something changed in the last year with Sam. Before that he used to be down with chasing women. He was wild, that being one of the few things we had in common. I could count on him to say yes to virtually everything, but over time, that stance changed. He spoke with more clairvoyance, exercising a special ability to see a life beyond drinking, liberated women, and adventure. He began to talk more about his desire for steady employment, a clean place to live, and a woman to share his life with. It was becoming clear that Sam was, *gulp*, maturing.

"I'll let you know if anything comes to fruition with this chick," I say before we get back on our bikes and continue.

We aren't far from the city of Lyons when we take another break. My mind is going crazy. I slam peanut butter crackers, and a chemical inside my brain is released when cell phone service crops up because now I can check Tinder.

"You're not looking for a relationship?" a mildly attractive woman asks when I invite her to meet up.

"I'm on a bike tour. I can't do a relationship."

"Fuck. I just wasted a superlike on you," she says.

Deleting Tinder would save me a few headaches, but I'm not ready to stop swiping and give up on what I want.

After we pull into the driveway of hosts Harold and Nora, it feels good to be off the bikes. After eighty-five miles, my legs are sore; and since we have another long pull tomorrow, I'm going to unapologetically eat the ass end of a boar pig tonight.

"I was just telling Sam," Nora begins while I stand in her kitchen after taking a shower. "You guys are going to have to leave by six thirty tomorrow morning."

My head starts to hurt.

"Maybe I could stretch it to six forty," she says with a smile that feels inappropriate.

I know it as soon as Sam shoots me that look, the one that is wondering if I concealed the early-morning wakeup from him. Had I known an early morning awaited, we wouldn't be standing in this house.

THE LONG ROAD EAST

In piling on more miles, sleep becomes more imperative, but those hours have been difficult to come by. Being out east, everything is more hurried. People have places to be, and gone are the days of playing on my phone for an hour before breakfast.

Nora's husband Harold instantly weirds me out when the four of us sit down for dinner. He doesn't offer Sam a beer, he's wearing overalls, and his eyes tell a story different from the words coming out of his mouth.

"I look at my neighbor every day. He mows his lawn, and he has a horse," Harold starts. He then looks at me like I can finish the story for him. "Anyway, he cuts his lawn and then buys food for his horse."

Everyone is still not following.

"His horse could eat the grass he is mowing!" Harold clamors, his voice now slightly below a yell.

My eyes are trying to plead ignorance, to no avail.

"And everyone else," Harold continues. "They mow their lawns, so they use gas purely for vanity. You don't need to mow your lawn, and you also don't need to use gas to do it," he says, going on to talk about a tool that cuts grass without oil.

Harold's message may be accurate, but his delivery is nothing short of R. L. Stine.

"That, that…that sounds interesting," I stutter, hoping he won't ask if I have ever mowed a lawn before.

I am still hungry after devouring the paleo dinner Nora had served. Going to bed hungry is horrible, but the sooner I get out of Harold's clutches, the better.

"Do you still want to see that chicken coop?" Nora asks Sam.

"Of course," he replies, leaving me alone with the eco-crazed Harold.

"You married, Quentin?" he asks.

"No, I'm not," I reply.

"Why?"

"I don't know if I'm going to get married. I mean, I might, but not right now."

Harold's face wrinkles with disappointment.

"Besides, it seems like a broken system. I have no interest in getting divorced," I say.

"How would you start a family?" Harold then asks.

"I guess, I would just start a family."

"Out of wedlock?"

"I'm not very religious, Harold, but truthfully I haven't thought that far ahead."

"If everyone did what you are proposing, we would have all these broken homes and unsolidified families."

"Just because people aren't married?"

"Yes."

"Well, I personally don't believe you have to be married to be a good parent. Look at all the divorced couples who make it work."

"Those kids are miserable," he says.

"Maybe. I don't really know. It's not something I've thought a lot about."

Harold looks at me with concern, our generational gap only fueling the molten fire embedded in his eyes.

"I'm going to go to bed," he says, then rises and walks out of the dining room without making any more eye contact.

CHAPTER 28

Nora drives off in her eco-friendly sedan as Sam and I prepare to leave for Ithaca. True to form, Harold is chopping away at the stringy grass in his lawn with a contraption that makes him look like a murderer.

"That guy is kind of weird," I tell Sam, getting off their property and never looking back.

"It's too early for this kind of talk," he mutters. "Just focus on what we need to do today."

Early on, the roads are getting more tumultuous, but Sam is still setting a great pace, once again stopping twenty miles in for lunch. From there we roll all the way into Ithaca with relative ease.

"Should we eat?" Sam wants to know when we touch down in the middle of an Ithaca town festival.

"Why don't we wait until we get settled? We are almost there," I say, noticing large crowds that would be hard to walk a bike through.

"But we are already here, and I'm hungry," he insists.

"Fine. I guess it doesn't matter."

We wade through the masses, Joad a nuisance as I maneuver her around the throng of people coming from every direction. There are kids everywhere, and it hits me at this moment that having kids right now would be horrible.

Forget Harold's talk on nuclear families. I simply don't want to deal with the crying, screaming, and bits of chicken hanging off my hypothetical child's lip.

"Danny, get back here," one mom yells after her child as he takes off with his shoes untied.

"It's so loud here," I tell Sam, following him as we methodically creep through the crowd before boxing ourselves into a corner.

Sam leaves to go find food, so I bury my head in my phone to avoid the concerned looks from passersby.

A cute girl from Tinder is insistent we meet, the catch being she does not want to drive two hours from Rochester to Ithaca.

"I biked here from Rochester. It's not that bad," I assure her.

"I'm kind of tired though," she sends back.

I roll my eyes, her earlier claim to have fallen in love after reading my blog seeming hyperbolized.

"I'm sure you can find some other coed to suck your dick tonight. You're at Cornell," she continues.

"But I want you to suck my dick."

"Aw, that's so sweet. I just don't know."

"I understand. No hard feelings," I tell her, closing the app and feeling frustrated because it shouldn't be this hard to get laid.

Sam soon comes back with some chili fries, and after eating them, we set out to finish the last few miles. The Warmshowers location is near the campus of Cornell, a university that sits atop a massive mountain.

As if the Ivy League institution was not already elite, its presence on top of this gruesome hill reinforces the desired effect. I give up pedaling when I come to a stoplight and start walking after three of my back muscles beg for mercy. The light turns red and then my phone dings.

"Terry and I aren't there," our host Jessica's text reads, "but we might be back tomorrow. Go check out some of the local fare."

I meet up with Sam on campus, and we bike the next five minutes together before arriving at a small duplex. It's not long before I'm in correspondence with a new woman on the Internet. She's not interesting or attractive, but she knows Sam and I have just arrived in town and is willing to take us to dinner.

"Sam, I've been talking to this woman on Tinder. She wants to go out to dinner with both of us. Would you want to go?"

"Are you trying to get me involved in a threesome again?"

"No. It's not like that. I don't even want go out with this woman, but it's something to do."

Sam looks at his phone and then stuffs it back inside his pocket. "I suppose I'm in. Not much else to do here."

The woman soon pulls up in her black sedan wearing academic glasses and sporting a smile that would scare off even the most audacious Jehovah's Witness.

"So you guys biked here?" she asks.

"Yes. We're from Minnesota," I say, a fact she couldn't be less impressed with.

"Tonight I'm taking you guys to an Ethiopian restaurant. It is BYOB, so I brought us a bottle of red wine."

"How sweet," I mention. "Unfortunately, I don't drink red."

"I like red," Sam predictably chimes in from the backseat.

"Of course you do, Sam," I quip.

The entire dinner turns out to be awful. The menu is a riddle, and the complimentary bread so doughy even a carb whore like myself feels violated. Two white guys in dreadlocks are walking around taking orders, and somewhere on the Internet a troll is screaming cultural appropriation.

Sam and our dinner partner, whose name I've by now forgotten, are hitting it off quite well, united by their shared love for alcohol. Despite their chemistry, I still don't understand the woman's rationale for coming out tonight. She doesn't want to talk about anything other than her lame job over at the college.

"At least I get good health benefits," she mentions at one point during her boring discourse.

Her being here is starting to feel like a last-ditch effort to fill up her Saturday night so she doesn't have to go back to work on Monday and tell the person a desk over that her weekend consisted of watching reruns of the *Gilmore Girls* and petting her three cats.

"How do you see the rest of the night shaping up? Maybe we could keep this night going back at our place," I say while the three of us leave the Ethiopian restaurant and walk back to her car.

"No. I'm going home," she says, flames firing from her mouth like she's Charizard.

"Probably for the best," I tell her, but I'm more so speaking to myself.

She drops us off, and Sam and I walk inside and sit down in the living room. There is a large map of New York state hanging on the wall that catches Sam's inquisitive gaze.

"Q, where are we going tomorrow?" he asks as we both stare at the map.

"Delancey, because otherwise we would have to do 128 miles," I sigh.

A dissatisfied Sam pushes back. "Q, look at this map. We are trying to get to New York City."

Sam's pointing at the state's epicenter that resides in the southeast corner.

"And we are here," he says, then motioning toward Ithaca, which is in the north.

"I can see the map, Sam."

"Then why are we going to Delancey?"

"It's just how the Warmshowers worked out."

"We are moving east, and we should be going south. We have to go back into Pennsylvania."

"Absolutely not," I state.

"Why?"

"Because we were already in Pennsylvania."

"But it's a more direct route."

"Sam, I'm in charge of finding places to stay. You're in charge of getting us there."

"Q, where you are taking us makes zero sense. And I'm putting my foot down on this one."

"Why? I never question you. Why are you doing this?"

"Because we are wasting time!"

"All we have is time!" I scream, angered by Sam's desire to take control. "Bro, you realize that once this trip is over, it's over. We can't just come back out here and keep going."

This is the first time all trip we have mentioned an end date. I hate Sam butting in, but now negative emotions whirl through my mind knowing this trip will soon unapologetically expire.

"You just never know what can happen. Staying east could be for the best," I mumble after Sam paces around the room to calm down.

"Whatever, man. This is clearly your show, and I'm just along for the ride."

Sam's victimizing language is insulting, but my poor planning put us here. Had I read a map and determined Buffalo and New York City are in different corners of the state, things would be different.

Sam and I will end up in Portland, Maine at some point. How we get there, and the status of our relationship, is still unknown.

CHAPTER 29

Heather, the Tinderella from yesterday who liked my blog, has changed her tune regarding coming to see me.

"I really want to meet you," she texts early the next morning.

"Why don't you just come out?" I ask, unable to fall back asleep because of the lingering tightness in my quads and soft rain splashing against the bedroom window.

"I'm thinking about it," she says.

"It's up to you," I finally write, hoping she throws caution to the Midwestern wind blowing in from the neighboring states.

A few minutes later she replies.

"Okay. I'm going to come. I'll be there in two hours," her message reads.

My heart rate spikes immediately. Forget hypotheticals. Now a plan is needed. As my mind goes crazy, a few foreign footsteps traipse past the door. They must belong to the hosts because Sam is still snoring. Their arrival is an indicator that I now need to devise an exit strategy to go meet Heather.

Confused and needing more oxygen in my brain, I walk downstairs into the kitchen. There stands Jessica, a marvelous beauty. I hope that when we shake hands, she can't see the awestruck in my eyes. A younger man then appears at the foot of the steps.

"Hi. I'm Terry, Jessica's boyfriend," he says.

I shake his hand and do my best to come off as submissive, thinking that will mask my salacious thoughts from minutes earlier. Sam soon joins us in the living room, and the clock says that Heather will be arriving in twenty-five minutes. My arms tighten, and I'm too excited for what's about to happen.

THE LONG ROAD EAST

Jessica conveniently leaves for work, which leaves only Terry, Sam, and I to converse in the living room. Raindrops continue to splash against the roof and exterior of the home. I look outside and see only gray skies and raindrops falling from the trees stationed next to the home.

My exit has to be done right. I need an airtight alibi to leave because transparency isn't an option. Near-guaranteed sex is minutes away.

Sam and Terry babble about politics, and I pretend to be interested, smiling periodically when one of them looks me in the eyes, but the truth is I don't care if Donald Trump is an asshole. I'm still living freely regardless of who the poster boy for our great nation is.

Ten more minutes pass, and I still have not concocted a fabrication for my departure from the home. I then let out a deep sigh and go outside and pretend to look at Joad, but really the fresh air is needed to clear my mind and come up with a reason for why I am going out in the middle of a rainstorm and not taking Sam with me.

I don't have many options if I want to make this work. I hate to do this to my best friend, but Sam is going to have to be kept in the dark. I can't have his moral righteousness sabotage my dalliance.

Once I get back inside, I feel a vibration coming from my right pocket. It's a text. Heather is here. Sam and Terry are thankfully still engaged in heavy conversation.

"Hey, guys, real quick. I'm going to go have breakfast with my friend. I'll be back in about an hour," I calmly interrupt.

Sam looks confused, so I nod at him and hope he gets the hint. Terry doesn't seem bothered and *holy shit* this is about to happen.

After descending the last step of their porch, I run toward Heather's car, a gray four-door that blends in seamlessly with the pouring rain. I race around to the passenger door and hop right in, thinking with nothing else but my God-given anatomy.

"How's it going?" I exclaim before lowering my head and sliding into the front seat.

Heather examines me, then smiles and puts the car in reverse. "Where do you want to eat?" she asks like I've thought that far ahead.

"I'm down with wherever," I say.

I whip out my phone and look for the closest thing to a Panera Bread. And then up pops an actual Panera Bread not too far away.

"Let's go to Panera," I suggest.

"Do you know where it is?"

"My GPS can get us there."

I look over and notice Heather isn't very tall. She stands a few inches over five feet, but she has an inviting smile. I can ride my bike between the gap in her thighs, so I'm in heaven, no matter how temporary my stay might be.

At the restaurant, Heather's words aren't resonating. Everything is a blur as I wolf down the cold ham and corn that cost $4.73 after tax. I'm cold from the overcast weather and feel slightly hungover, even though I haven't consumed alcohol in days. A glass of orange juice gives me a boost, and I can't pretend this morning is about anything other than lust.

"We should get out of here," I say after checking my phone and seeing that time is slipping away.

"Where do you want to go?" Heather asks as she stands up from her chair.

"We'll find someplace," I confidently assure her.

As we walk to her car, eerily cold rain continues to fall from the sky, the chilly wind bringing me as much discomfort as the thought of Sam soon pestering me with phone calls. He and I still have a long ride ahead of us today. He also has no idea where I am and is in many ways stranded without breakfast while sitting on the couch of a complete stranger. My whole stunt is selfish, but in no universe would I turn this opportunity down.

Heather whips her car behind a strip mall, parking a couple hundred yards away from a dumpy motel that is barely visible through the pouring rain.

"This okay?" Heather asks.

My head cocks both directions, paranoid the cops stalked my Tinder account and are aware of my every move.

"This is perfect," I nervously and excitedly say, looking around the car for a functional and penetrable position.

THE LONG ROAD EAST

The backseat contains a stash of magazines and other miscellaneous items placed in a brown box. I extend my arm and push everything onto the floor, then awkwardly contort into the backseat and begin taking off my clothes. Heather does the same.

It's instant chemistry. Her lips clash with mine until she goes down on me for a couple minutes, the warm breath trapped inside her mouth gently consoling a part of my body that's been dormant for weeks.

"Do you have a condom?" she asks.

I chuckle. There are condoms in almost every pocket of my jacket, including spares stored away back with Joad.

"Of course," I say, and then she watches in excitement as I roll one on.

Heather sits on top of me, my entire body becoming numb. I sit comfortably for the next couple minutes as she rides me, peering out the window at the pouring rain, feeling at ease for the first time in weeks. The trip and all its stresses disappear for a few minutes. It's one of those rare occasions where I am in a moment rather than looking forward to another one.

More time passes and then inexplicably a switch in my brain is triggered. Joy begins to fade. Now I'm concerned with everything. The digital clock above the air vents is a few ticks from eleven thirty. Time is not on my side.

Sex begins to morph from romantic to transactional. I would like to stay here all day, but circumstances will not allow that to happen. My eyes become glossy; the pent-up sexual energy harbored since Wisconsin being exerted. And then Heather puts her hands on my stomach and moans.

Numbers on the digital clock continue to change. I can't get out of my own head. I slide Heather off and pin her to the driver's side door. Her eyes are flush with pleasure, and now I'm a few strokes away from jettisoning what I think is important into a shaft of latex that protects my future self from regret.

I look into Heather's eyes when I can't hold back any longer and let out a moan. My heart is thumping, and my breathing is taking its time returning to normalcy. Motionless, my arm is still braced against the door so I don't fall on top of Heather.

The rain that continues to fall is soothing, so much that I want to collapse and never wake up, but reality keeps reminding me Sam and I still have sixty-plus miles to ride. The nagging clock on the dashboard catches my eye again, and I frustratingly maneuver to the other side of the car.

Heather isn't done.

"Just finger me a little bit," she says.

She grabs my hand and leads the way.

I continue until she grabs my wrist while she orgasms, but instead of feeling happy I am relieved she's finished so we can head back. I keep looking at that damn clock, as if by doing so time will rewind and bring me back to the only place I can stand to be.

Heather has existed in my reality for less than an hour. The emotional hole I now possess quickly erases the physical satisfaction, our liaison now stuffed into a file folder no key can open. The moment is gone and won't ever be recaptured.

"I wish we could spend the day together. Just lie in bed and keep having sex," Heather says, moving to the front seat and turning on the car.

Guilt invades my headspace. Sticking around would be awesome, but the road ahead is calling. We drive back to the house, and I ready to say goodbye.

"I'll see you later," Heather says.

I give her a hug and then exit the car. "I hope so," I say.

I don't have the guts to tell her we will never lock eyes again, that what just happened was a moment in time that won't be repeated. It isn't because I'm that douchey guy who got what he wanted. Circumstances dictate reality, and I can't imagine a situation that will ever bring me back to upstate New York.

I shut the door and begin blocking out emotions. Regardless of the joyous feelings I have, that's all they will ever be, merely images replaying in my head.

I casually walk inside.

"You ready to leave?" I ask Sam as if I had simply walked down to the gas station on the corner.

"I've been ready," he says coldly.

We say goodbye to Terry and then convene outside under the pouring rain.

"Did you get laid?" Sam asks.

I coyly smile and throw back my shoulders. "Yes, I did."

Sam doesn't say anything back. We head out into an unforgiving downpour, Ithaca simply a thing of the past.

CHAPTER 30

"That guy was so fucking annoying," Sam ripens the next morning with after we leave Binghamton, New York.

"Who? The host Glenn?"

"Yeah. Typical fucking Democrat and Bernie Sanders supporter. Wants everything to be free and…"

Sam makes laughing easy.

"I can see why his wife left him," Sam continues, enough salt spewing from his mouth for a small country to mine.

"Damn, dude. I thought he was an okay guy," I pipe in. "He did give us free blueberry oatmeal."

"Yeah, and it tasted like shit," Sam sneers.

"Sam, I think this is the first person all trip you *haven't* liked."

"For good reason. Even when you were talking, he'd cut you off and say the most meaningless bullshit."

It's hard to stop laughing because not only are Sam's barbs hilarious, it's also nice to not have to be the bad guy for once.

Sam has to momentarily suspend bashing Glenn when not even five miles out of Binghamton we encounter a monstrous hill. Beads of sweat coarse down my breastbone as I labor up the daunting road. My muscles are stiff and not ready for this early challenge.

Another hill follows, and then another. Going down is enjoyable, but going up each new hill is more difficult than the last. A short break comes when we ride downhill for an entire mile, going so fast stopping at the bottom requires me to squeeze both handbrakes with all the strength I can muster from my measly wrists.

"No fucking warm-up either," Sam bitches as we stop early for lunch on the side of a highway.

THE LONG ROAD EAST

Based off the early terrain, it's going to be a long night. It's noon and we still have sixty more miles to go.

"They're just hills, bro. We can grind through them," I encourage.

"Such a fucking joke though," Sam continues to bemoan.

I take a handful of chocolate chip cookies from my bag and begin stuffing them in my mouth.

"Go easy on those, Junior. You don't want to get diabetes," Sam jokes.

"Your fat ass is closer to being diabetic than I am," I fire back.

"You wish. I'm in great shape," he says, grabbing his belly with both hands to emphasize his point.

"But seriously though, how have you not lost weight on this trip?" I ask. "I've lost eight pounds."

"It's a mystery," Sam laughs because he's so comfortable in his skin that no matter what he looked like, Sam would still be the most charismatic guy in a room.

"Evidently."

Once the cookies are finished, I reach into my bag in search of anything that can be consumed. Ironically, the more fistfuls of food that go down my gullet, the hungrier my stomach becomes. There must be a science to explain this madness, how my body can be in constant need of replenishment.

"Fuck, I can't stop eating," I tell Sam.

"Yeah, you can. Just stop. It's that easy."

I laugh. "When did you become the expert on responsible eating?"

"Just now," Sam smiles.

My package of cookies: gone. The Pop-Tarts: gone. It shouldn't be possible to eat this much and still lose weight.

Cars cruise by, and the passengers inside stare as I shove the wrappers in my bag. It's so hot I shed layers until the sun is free to burn parts of my skin that haven't already been scarred. I check my body and notice a bad tan line has surfaced midway up my arms.

"Take a photo of me," I request, hoping my abs will look great.

Sam takes a photo, and when I look at the results, the camera has captured a cross between a redneck and a Holocaust victim. My farmer's tan and the noticeable lack of bulging abs is demoralizing.

"You're a fucking twig," Sam maniacally giggles.

"It's because I don't have my protein powder, asshole."

"Right. Keep telling yourself that, Q."

The tenor of our conversations has been dicey ever since I left Sam in the home of a random stranger. Knowing us, we won't address what I did until one of us can't hold the anger in any longer.

After lunch, the riding doesn't get better. The constant hills are more irksome than rush hour traffic. It's drizzling rain and nearing dinnertime when we pull into a small town twenty miles from Delancey, New York, craving something other than oatmeal cream pies and chocolate milk.

"Are you guys wanting to eat here?" a dark-haired man asks as we disassemble outside his pizza shop.

I look at my phone, and it isn't even six o'clock.

"Is that okay?" I ask.

"We're almost closed, but I'll stay open if you guys are hungry."

"I'd definitely like to eat, if you're willing to stay open."

The man looks back inside for a brief moment. "Sure, come in," he says.

Inside, the restaurant has the feel of the bar from the TV show *It's Always Sunny in Philadelphia*, which ironically is playing on an old TV on the other side of the room. There isn't much life to the establishment. A few tables are lined up against the windows, with a big Coca-Cola fridge placed near the cash register. There are also no photos or paint on the walls, suggesting this place has not yet finished its construction process.

"Enjoy, fellas," the owner says after he brings out a large sausage and pepperoni.

I don't want to go back out into the rain and continue trudging up mountainous roads. We still have twenty miles left, and if the terrain continues at this pace it will be at least another three hours before the ride is over. My gut feeling is that the roads won't flatten

THE LONG ROAD EAST

out, but I suffocate those doubts by stuffing slices of pizza down my throat until vomiting seems consequential.

"Where are you guys headed?" the owner asks.

"Tonight, Delancey. Eventually, Maine," I tell him.

"Sweet, sweet. You guys want to do a photo for Facebook? I want to promote the store."

His family is waiting outside in a minivan. His question feels like a polite way of asking us to leave, and since my headstrong opinion doesn't matter, the three of us stand close together as the chef readies to snap a photo.

It's hard to muster a genuine smile. I don't feel welcome in this man's restaurant, so instead of politely explaining myself, I clench my teeth together and force my cheeks to raise, giving off the impression that feelings of joy occupy my current headspace.

"Best of luck, fellas," the man says after figuratively pushing us out the door. He then hops in his van and drives away.

The rain has intensified, and darkness is an hour away. Half an hour later I piss on the side of the road knowing full well a pickup truck might crash into me and end the utter annoyance that is my life. Getting run over seems better than subjecting my brain and lumbar to the excruciating road ahead. The joy of this trip is quickly dwindling.

I take out my phone to give our host George an update. "We won't arrive until late," I tell him.

"How far away are you?" he asks.

"Probably fifteen miles, but it's all hills, George."

"You're in the Catskill Mountains. What did you expect?"

"I didn't have any expectations, George. We'll see you soon."

A break comes a few miles later when the rain lessens to a drizzle. Handlebar lights are our only savior since the evening has now turned to night. I can't see much of the road ahead, even as we begin dropping down a long decline.

It doesn't matter that visibility is threatened. Sam can put us in a ditch for all I care. I'm more impressed by his unwavering fidelity to strong-arming us through this prolonged trek.

Two miles from the destination, another snag. It's a skyscraping hill, extending all the way to the address.

"Just keep the pedals moving," I tell myself while wrenching my bike back and forth to generate momentum.

My lumbar is in a vise grip, the muscles stretching and begging for relief with every stroke of the pedal. I want to start walking, but doing so would be cheating Sammy and all the effort he gave to drag my sorry ass this far. Believing today will end won't happen until my head hits a pillow, or the pavement.

After fifteen minutes of lugging Joad up this hill with whatever strength I have left, the turn mercifully arrives. I get off my bike to stretch my back out and thank a deity it's over, but I spoke too soon.

George's address is at the base of a hill. Trees protected us from heavy rain the last two miles, but now we are back in the open. Before the biking gods grant us a pardon and deliver us our salvation, we have to ride onto a gravel road full of loose, slippery rocks and up another steep hill.

"How are we supposed to ride on this stuff?" I turn and ask Sam.

He has no answer. Instead he forges forward. His tires begin to slip on the loose rocks and then he dismounts and begins walking.

"I don't know how to explain what happened today," he admits.

I push past him while walking up the hill and come to a mailbox. "Finally!" I yell back to Sam.

We're not quite done yet though. George's house is down another small hill. I begin to slowly creep down the slick, muddy driveway.

Joad then turns into a raging bull, pushing forward while I dig my heels into the soft mud and attempt to keep her at bay. She's going to run me over if one of my shoes gives out.

The faithful Joad has every right to trample me. I have asked a lot of her and not given much in return. Her manslaughter seems more humane than the eighteen-wheel semi back in Michigan that at the time had no stake in my future.

A part of me wants to let go of the handlebars and crumble. It seems easier than subjecting my body to more labor. I'll lay in the rain all night and let ants crawl in my mouth if it means I can go detach myself from an apparatus that's taken me all over the land of the free and the home of the brave. Bury me with my name in the sand. This is where I'll be happy to lay down for good one day.

The pain ends when we reach the house. A small light is visible inside the cabin. It looks warm inside. George and his friend are huddled under an awning on the wooden porch.

"Put your bikes over here," George yells through the pouring rain.

We shake hands, and he asks me how we got this far.

"We have nothing better to do," I tell him.

After orchestrating a hot shower and some pasta, George and his friend Fred prove to be an intellectual bunch. Avid readers, they are interested in my soon-to-be-released book, *The Long Road North*.

"It took me two years to write the book, although believe it or not, the weather on that trip was worse than this one," I explain.

"I'm proud of you for writing it," George says. "It's hard to will yourself to finish."

"What kind of books do you like to read?" Fred asks. "George and I read a lot of books."

They're looking at me expecting a dazzling response.

"Listen, fellas, I'm not the most well-read guy in the world," I tell them. "But I do like to read memoirs. A few months back I read *Wild* by Cheryl Strayed. And then, I can't remember the author, but I really liked a book called *This Boy's Life*."

"Fred, who wrote that one?" George asks.

"Tobias Wolff," Fred answers.

"Might have to read that one," George says, then turns to Sam. "And what about you, Sam? What kind of books do you like?"

"I read a lot of Dan Brown a few years ago," Sam replies.

I quietly sneak off to the shower while Sam has a beer and discusses the finer points of *The Da Vinci Code*.

Later that night I find myself spread out on a grimy pullout mattress in the living room, the fire from George's woodstove a neutralizer to the evening's punishment. The dog hair and brown blotches on my pillow simply cannot deter the exhaustion I am feeling. I did a lot of searching today, but looking for a clean pillowcase was not one of them.

CHAPTER 31

Sam has leftover clothes still sitting in the dryer. I need to run mine through, so I take his out and set them on the floor. This will irritate him, not because his clothes are on the floor, but because the floor at George's place is littered with his dog Buster's hair. The devil inside me laughs when Sam goes to pick up his clothes. I can see his nose wrinkle and his face turn despondent.

"Did you have to throw my clothes on the floor?" Sam asks when he comes into the living room.

I pretend to be oblivious to his qualm. "I needed to get mine in. I just set them next to the dryer. Are some missing?" I ask, my question diverting from the core issue.

"No. It's just…Uh, never mind," Sam replies with utter frustration, unable to see I am lying through a pair of teeth that only look nice because my parents got me braces when I was in eighth grade.

I should have asked Sam to take his clothes out or put them somewhere safe from Buster, but doing so would have required a selflessness I can't seem to offer.

"If I knew you would have been mad, I wouldn't have done that," I continue pushing.

"Let's just get out of here before I go nuts," Sam says.

The day ahead awaits, but I can't find my phone. It's not in the bathroom, laundry room, or my bags.

"I'll call the number," George says, and then a faint ring comes from the pullout couch.

"Maybe it's under the couch?" I ask George.

"Let's just move this thing out of the way," George chirps, a task not easily done because the couch is heavy.

THE LONG ROAD EAST

We work in tandem to move the thick wooden couch, but my lumbar still aches. My phone continues to ring at a small decibel.

"Is there a black hole under here?" I ask George.

The more we move the couch, the worse the floor looks, and I'm hoping my phone is not stuck in one of these mounds of garbage that are smooshed together.

We shift the couch another way and there is my phone, muddled with all sorts of nastiness. Numerous dog bones, wood chips, wrappers, dust, and general abomination lay on top of my lifeline.

With reticence I pick the phone up and slide it into my pocket, trying to remove the nauseating visual from my consciousness as quickly as possible.

"It looks okay," George states.

"Something like that," I reply, and then Sam and I head back out into the Catskills.

"I didn't mean to freak out earlier. That place was just so disgusting," Sam says later on while my legs pedal lackadaisically. "I mean, do you think George ever had guests before?"

"I'd imagine we weren't the first people to stay at his house, Sam."

"Sure, but that place was horrible. I bet it started after his wife left him."

I chuckle, relieved Sam is pissed at someone other than me or Bernie Sanders supporters.

The hills of upstate New York continue unrelentingly. Rain says hello like it's an old friend. If this is karma for my laundry stunt, it's totally justified.

The weather in Michigan was dreadful, but upstate New York is equally macabre. My tires don't grip the road well, and most days the back of my jacket has gotten doused in dirt and grime. I could purchase fenders to prevent the dirt from moving up my back, but Sam is constantly fiddling with his after crashing his bike during a tango with Blue Moon, and I don't want to deal with that same annoyance.

We stop at a gas station around lunchtime. Along with a few slices of pizza, I suck down an entire half gallon of chocolate milk from a glass container, and within ten minutes I feel more bloated than a pregnant woman.

Outside the window, I see a duet of retirees looking at our bikes. Their eyes sparkle when they point at Sam's bike, and then one of them grimaces and laughs when they point at mine.

The man then turns and spots us through the window. Him and his wife hustle inside to say hello.

"Are those your bikes out there?" the man asks upon approach.

"They are," Sam responds excitedly.

"Who has the Salsa?"

"I do," Sam's quick to point out.

"That thing is beautiful."

The man glances at me and realizes the other bike belongs to me, then deftly segues onto another topic.

"Anyway, do you guys have a place to stay tonight?" he asks.

"We do," I respond. "In Phoenicia."

"I'm going to run over to the grocery store real quick, honey," the woman says to her husband.

"Okay, sounds good, sweetie. See you outside," the man says. "So anyways, guys, I was biking through Wyoming," he then narrates without solicitation.

My stomach is expanding further than the strike zone of a minor league umpire, and listening to this homer talk about his past while he exclusively makes eye contact with Sam makes me want to stand up and leave.

"We got to get going," I say, interrupting the man's monologue after five minutes.

I walk outside and ready for the final stretch while Sam finishes the conversation.

"Bye now!" the man waves excitedly at Sam as the two walk out of the store together.

When we get to Phoenicia a couple hours later, having survived a plethora of hills and the ill-fated Google Maps, we stumble upon a neighborhood nestled in the woods. It's positioned only a block from the highway.

THE LONG ROAD EAST

It's there we meet Carol. As she washes vegetables before dinner, we learn she wrote a book documenting her life as a sex worker. I've never encountered a woman who used to be in that industry, so it's hard to find the right things to say.

"The women who do this for a living, are they empowered?" I ask.

"Some of them," Carol says. "Others do it to escape bad situations."

As our conversation evolves, Sam slips outside to avoid the awkwardness pervading the room.

"Where you going?" I ask him.

"Just need to check on a few things," he says back.

"I think he might be a little uncomfortable," I tell Carol.

"Are you uncomfortable?" Carol asks.

"No, but I do want to be sensitive to what we're talking about," I say.

"Sure, but there's no need to be apprehensive. If you want to know something, just ask."

"I'll keep that in mind," I say as Sam walks back into the room.

Over time, the night begins to age. Sam and I aren't threatening, but we are in our twenties, opening up the real possibility that body language and words can be misinterpreted.

"I apologize if I'm being invasive," I tell Carol after asking another question. "I've just never met a sex worker before."

She grabs another book from the shelf, this one a collection of stories from other women in the industry. "You're not being invasive. I don't mind talking about my past," she reassures.

Carol hands me the book, and as I page through it, my eyes begin to flicker. It's time for bed. Sam is sitting on the couch opposite me with a pale ale in his hand.

"You look sleepy," he says.

"Long day on the road?" Carol asks.

"The Catskills have drained me," I admit. "I think I need to go to bed, guys. My eyes can't stay open much longer," I tell them, then head for my bedroom before curiosity gets the best of me.

CHAPTER 32

"Carol was different," I say to Sam as we begin the slow ride back to the highway.

"She was."

"I knew you were uncomfortable when you left to go check on your bags in the pouring rain."

"No. I wasn't uncomfortable. I actually had to go check on something."

"Right. In the pouring rain. Sure, bro."

"Q, I did."

"So you weren't turned off by the conversation?"

"No."

Sam has this way of tactfully evading situations he wants no part of. He'll slither out of a room and go elsewhere, resigned to the fact that a certain topic isn't for him. He thinks people don't notice, but with him being the centerpiece of every conversation, it's very obvious when a social butterfly like him exits the room.

"Whatever you say, Sam. My point isn't even important. I just wanted you to admit you were uncomfortable."

"There was one thing though," Sam then says.

"What's that?"

"I do think she was wondering, 'okay, which one of these assholes is going to try to fuck me?'"

And then I laugh. "It definitely would have been you. You haven't gotten your dick wet in a minute."

"Shut the fuck up, kid," Sam caterwauls. "Just focus on the road."

Courtesy of my suspect orienteering, we are still stuck in the Catskill Mountains as we try to cut down into New York City.

THE LONG ROAD EAST

"I just wanted to say I climbed one mountain. Not every mountain," Sam bitches as we rest on the shoulder of a highway.

"If you want to be mad at me, I won't blame you," I respond.

I messed up. Being this far up in the mountains makes about as much sense as investing in alcoholic sodas. Life certainly would have been easier had we dropped down back into Pennsylvania, and even worse, this route takes us farther from Portland, Maine.

Sam continues to yammer away, and I reach into my bag for another packet of cookies when suddenly an SUV pulls up behind us.

"Are you guys okay?" the man asks as he gets out of his car.

"We're fine. Just taking a break," I tell him.

"I saw you guys parked on the side of the road, and I just wanted to make sure you weren't in trouble."

"That's nice of you, but we're good."

"I own a bike shop in the city," the man then transitions. "You guys headed that way?"

"We are. Going to spend some time in the Big Apple," I reveal.

"Ever been?"

"No. This will be our first time."

"The city is wild, man. Absolutely wild," he says, his eyes lighting up.

"And the women?" I ask.

"Shit, they're crazy, man. Be careful."

He says goodbye and then we ride further toward the city of Marlboro. With each passing mile, the traffic and noise become louder. The call of America's greatest city is beckoning us.

A short while later we get turned around and somehow end up inside a small forest.

"Why the fuck are we in here?" I yell at Sam. "You trying to mountain bike or something?"

"Just chill, dude. The GPS is all fucked-up."

"Real surprise there."

We ride further along a trail filled with tree branches and mosquitoes before arriving on a circular field of grass that's a dead end.

"What the fuck! This shit is so whack," Sam squeals, using his index finger and thumb to probe deeper into the map on his phone.

"What about this bridge?" I say.

"What about it?"

"We could just walk through this tall grass and connect with that road up top."

Sam surprisingly finds my suggestion plausible, so we drag our bikes up the hillside, and then mercifully are granted access to a trail.

"I just need to find out which way to go on the trail," Sam says, turning in half circles in an attempt to orient his GPS.

"Ask that chick," I say, noting a middle-aged woman walking toward us.

"Why do I have to talk to her?"

"You're good with strangers."

Sam puts his hand up, and the woman stops to address him.

"Excuse me, miss," Sam says politely.

"Can I help you?" she asks.

"I'm hoping you can," Sam says, but the woman is not helpful, proving to be a bit touched in the membrane the more she speaks.

"I think it's that way. Wait, I think it's that way," she stammers.

"If you don't know where it is, that's fine, miss," Sam tells her.

"Oh, I think I know. Just give me a second."

The normally hospitable Sam is growing impatient. "Okay, miss," he finally interrupts. "I think we can figure it out from here."

"Oh my, I'm so sorry. I've just had a lot of problems. You see...," the woman begins, at which point I start walking the opposite direction.

Sam hears her out for thirty more seconds before disconnecting from her cringey soliloquy.

"Some people," he whines after catching up.

"I didn't want any part of that chick," I say.

Later that afternoon, our destination is nearing. We pull over on a side street to reflect upon another day filled with tumultuous roads and bizarre characters.

"I could eat the ass end of a boar pig," I tell Sam.

"Of course you could, kid," Sam replies.

THE LONG ROAD EAST

A truck then peels around the corner. I assume the driver sees us idling on the side of the road and will adjust, but instead his front wheel nearly clips my left foot while the diesel engine roars with fury.

"What the fuck!" Sam screams after the car.

The man slams on the brakes, directly parallel from me.

"What the fuck did you say to me?" the driver yells out the window.

"Wait, what the fuck did this idiot say?" Sam shouts, then marches right toward the passenger window.

"You got something to say?" the driver yells back.

Sam gets to the window.

"What the fuck, asshole? Are you trying to hit us?" Sam demands to know.

"You're in the middle of the road!" the driver yelps.

"So? Does that mean you're just going to run us over?"

"I have a right to be in the road."

"So do I!" Sam barks.

The two trade barbs amid rising tensions. The man has the support of his diesel engine, but staying seated is his best play. You don't want to fight Sam.

I remember there were a few times I wanted to punch Sam's teeth in, like when he blabbed about the fact that I never played high school basketball.

"You're a gym class hero!" he kept yelling in a drunken stupor one night.

It took a lot of energy not to march over and begin swinging, but real talk, Sam would kick the living shit out of me if we ever got into an altercation. If it wasn't his massive belly and forty-pound advantage that would do me in, it certainly is the fact that in high school he was one of the best wrestlers in the state of South Dakota.

Sam came from a small town where there was a lot of farming. You don't mess with kids who lift haystacks and milk cow udders for fun. They are a different kind of strong, a strength us suburban kids in skinny jeans will never possess.

But the man operating this truck does not know of Sam's potential.

"Get the hell out of the road!" the man demands once more.

"Or what? You're going to run us over?" Sam screams again.

I can stand idly no more. Only something bad can happen at this point.

"We were wrong," I say to the man. "We were in the road."

He mishears. "What! *I* was—"

"No!" I yell. "*We* were wrong."

He finally shuts up and lets my words sink in.

"Okay," he says, and then quickly drives off while making sure his engine is heard by everyone within a three-mile radius.

"What a fucking douche," Sam jaws.

"Just let it go. We're almost there," I say.

Tonight, our Warmshowers host is a lovely woman in her golden years. She and her husband own a winery that overlooks an expansive forest.

"You don't have to sleep outside," she tells us. "I only said that earlier in case you guys were scammers."

"Scammers?" I ask.

"You know, someone who just wants a free place to stay."

"We're legit," I tell her. "We have come all the way from Minnesota."

"I believe you. It's just one time I had this fat guy show up on a bike. Didn't look like he had biked more than a mile. I mean, he was fat, and he wasn't at all sweaty. He didn't have any gear either."

The woman's brutal honesty is refreshing.

"And then he wanted to stay the whole weekend. *And then* he wanted to come back the next weekend! Geez, that guy was a fucking idiot," she laments.

"Well, I can assure you I'm not fat," I joke.

For dinner, Sam and I walk to a highly recommended restaurant in town. The specials are thirty-dollar plates, and when Sam orders the first of what will be multiple Long Islands, I imagine the bill will be massive.

"Do you guys have chocolate milk?" I ask the woman behind the bar.

"Yeah, we do," she says.

"Is it actual chocolate milk, or do you just put chocolate sauce in the glass?"

"We use the sauce."

"I only ask because I need the protein."

"That's great. Do you want the chocolate milk or not?" she then asks.

"Good question. I need to think about it."

The woman rolls her eyes and walks away.

This restaurant is expensive, but it's also good. My lemon chicken and risotto are divine, and if I leave here having spent under forty dollars, it's a win.

"I'm kind of pissed," Sam says when he gets his bill.

"Why?"

"My bill is almost seventy dollars before tip."

"You did have three Long Islands."

"I just wish they wouldn't have told us to go to the most expensive place in town," he says, now decrying our Warmshowers hosts for their recommendation.

Sam's logic makes sense, but it's a compliment our hosts didn't take one look at us and say Burger King was down the street.

We aren't in trouble financially, but the cost of living keeps rising the further east we move. We were living like kings in the Midwest, surviving on ten-dollar days. Then at the first gas station in New York, I spent over twenty dollars on lunch and a bottle of Listerine.

Nothing is going to change the reality that life is now more expensive. We simply have to be more frugal.

Back at the hosts' winery, Sam is dabbling in the house blend as everyone sits outside and admires the majestic nature best viewed from lawn chairs. Fortunately, the wine and the conversation soon revive Sam's foul mood.

"Feeling better?" I ask the gregarious Sam.

"I was never feeling bad," he says, then goes back to speaking with the husband.

With LeBron James and the Cavs in the finals, I scramble upstairs to watch TV and stake claim to the spacious bed Sam won't want to share.

"So I'm guessing you get the bed," he asks midway through the third quarter as he posts up on the couch.

"We can share it," I offer, knowing he won't accept.

"Just sleep there, Junior," he says.

At the end of the game, Kevin Durant hits a three in LeBron's face to take the lead in the final minute, and this ends my dream of seeing LeBron win another NBA title. I die a little bit when Durant's shot goes through the net, and not only because basketball season is coming to a close. KD's shot is a reminder that soon these nights with Sam will also belong to history.

CHAPTER 33

New York City is getting closer. The roar from the millions of people who inhabit the city swims up my tires and jets through my legs. NYC was always a place on a map, but now it's a place I will get to see with my own eyes.

Right before the suburb of Tarrytown, we climb narrow side roads with no shoulder. We look like idiots, moving at a snail's pace lugging gear up these winding and rolling hills. Drivers who just got off work move over just enough to avoid clipping our back tires.

It isn't uncommon to feel pinched, but these roads are too narrow. I grab my handlebars and hope a sideswipe doesn't knock me off my moorings. I put Joad out further into the lane of traffic to make cars adjust and give space, but there is simply no room for bikers on this road.

Sam is fatigued when I pass him on an incline, and not long after that I am flying down the backside of a hill. A three-way intersection is at the bottom, so I pull over and wait.

After fifteen minutes of waiting for Sam, I take out my phone.

"Where are you?" I ask after he answers.

"You're going to have to come back. Our turn is up here," Sam mentions.

"You serious?"

"Yeah."

"Fuck me."

Frustrated, I hang up the phone and ride back up the monster hill, cars now moving faster as time sneaks deeper into rush hour and the volume of vehicles increases. After a few minutes, I see Sam off to the side of the road in a residential neighborhood. I need to make a left turn to get there, but waiting for a gap in this mess might take ten minutes. A kind soul stops and then someone parallel does as well.

"Thank you," I say, throwing up my hand as a goodwill gesture.

We meet Daisy and her British husband outside their home that parallels the Hudson River. We're staying in the basement, which is nicer than most hotel rooms. It has a bedroom, living room, luxury bathroom, and in the fridge a six-pack of pink soda.

"You want one?" I ask Sam.

"Nah, I'm good."

It only takes me thirty minutes before half the pack is floating around my stomach. A quick glance out the window reveals we are a thousand yards away from where Sully crash-landed that plane a decade ago.

"We're just going to kind of leave you to it," Daisy tells us when we go upstairs after showering. Her tone is polite, but I can tell she doesn't want us around.

This means Sam and I won't have to share our heavily practiced introductions regarding where we're from, what college we went to, and if we have girlfriends.

Back to my phone I go. A woman named Macy has shown a passive interest in meeting after I explain our trip and how I got to New York.

"I think we matched yesterday, but I'm in Tarrytown now. How far away is that from you?" I ask.

"Two hours," she responds.

"Bit far, but if you're willing to drive out here, it'd be fun to meet you," I send back.

"And you said you're staying in someone's house?"

"Yes. We have the downstairs to ourselves. It's just me and Sam, my riding partner."

She openly discusses the pros and cons of coming. "Let me think about it real quick. I'll call you later," she says.

Sam and I hit downtown Tarrytown for dinner, stopping off at a barbecue spot.

"This woman I've been talking to is probably coming over," I tell him in between bites of brisket.

"What do you mean?" he questions.

"A woman I know will likely be coming to the Warmshowers tonight," I clarify.

"From Tinder?"

"Yeah."

"Why, Q? Just why?" he groans.

"What do you mean?"

"Never mind."

I'm turned off by his tone. Our dialogue ends, and the disconnect between us continues after dinner. We can't agree on a place to go have a drink, and Macy has now confirmed she is coming. Once again, I have to balance my prerogative with the needs of my best friend. And by balance, I mean disregard Sam's wants and desires in favor of my own.

"Well, what do you want to do?" I ask Sam, sloppy indecisiveness hanging over us as we walk the streets.

"I don't know," his mumbling words so unenthusiastic I want to leave so he can binge drink by himself.

"You seem to have a problem with everything I do," I tell him.

"Huh?"

"You heard what I said. Have my fucking back every once in a while, bro."

Sam exhales. "Dude, I don't even know what to say to you. We both know you're going to do what you're going to do."

This discord has lasted a week, during which Sam has grown increasingly irritable. I can't pinpoint why he's upset, and he isn't telling me. His thoughts are more unreachable than cell phone service in a desolate basement. He's never been an emotional gold mine, but even this is unchartered territory.

For all I know, Sam ran over a cat a couple weeks ago and can't shake the guilty feelings that came when the animal hissed and cried while its tail took a beating. But the more likely answer is he's had it with all my chirping and antics, and the quicker we get to Maine, the better.

Darkness engulfs the Hudson River when we return. The light in the living room isn't inviting. Anxiety and excitement flood my head. Maybe Macy coming over is immoral, but just like in Ithaca, I'm not selfless enough to put a stop to what is about to happen.

Sam takes a six-pack of beer and heads outside onto the deck overlooking the river. He's talking on the phone when Macy arrives. I quietly go outside to greet her, and then we walk down the cemented steps into the basement. I've now likely violated Warmshowers' policies by having a guest over.

"How's it going tonight?" I ask Macy.

"I'm fine," she says.

Macy is a few years older than me, has dark hair, a warm smile, and nice legs. Everything feels like a blur as we sit on a futon, each of us occasionally taking the time to nervously smile and look away.

We relate on many subjects, biking being our most shared connection because she has also gone on a long-distance trip. A vulnerability develops that wasn't there while I gallivanted in Ithaca. As Macy and I talk, the more her being in this basement seems appropriate and less transactional, that this setup isn't as perverse as it was initially intended to be.

"You know what's sad though?" she says in regard to her trip.

"What's that?"

"When the trip was over and my boyfriend left, I cried for days. I couldn't stand being back in reality. I missed the journey."

Her words resonate with a stabbing profoundness, fast-forwarding me to a reality that grows closer by the day. The more she speaks the more all the equity built on this trip needs to be cashed in. It's painful to acknowledge that after the next few weeks are finished, only memories will remain.

I won't be able to feel the road beneath my tires in the same way, or joke with Sam in the same context. Every sentence will begin with "remember when," and we will acknowledge that we can think back to an earlier time in our lives, but we will never truly be there because the moment will have been lost to the past.

I don't know what will happen when Sam and I hit Portland. Maybe it will be the best day of my life, or maybe my heart will break, and I'll spend the next week crying and wishing Sam would yell at me over my reckless Tinder escapades.

"What are you thinking about?" Macy asks me, catching me staring out the window.

I laugh. "I've kind of run out of things to say," I admit.

We both know what's next. She likes me, but there is no empirical evidence she's down besides the fact she's seen the futon and hasn't dipped out yet.

"Do you want to go to bed?" I politely ask, embarrassingly motioning at the futon we are seated on.

"I think we're already in bed," she says.

"Yeah, um, that's my bad," I nervously chuckle. "Another thing is, I actually don't know how to turn this futon into a bed."

Macy looks at me and then at the futon. "Stand up," she says.

She grabs the futon and begins changing it into a bed. I move over to the wall to shut the lights off. We lay down on the bed and slowly things turn more intimate.

Real emotions circulate my brain as we begin having sex. Tinder brought us together, but fate is taking over now. I start dreaming of a day when she and I can take a trip together, and how I can be there for her so she doesn't have to cry when it's over.

Time has elapsed and then Sam walks inside from the deck.

"Huh?" he blurts, judgment seeping out of his beak like a leaky pipe.

I don't say anything. Seconds pass as I can sense he's standing and watching. "You're just going to keep going?" he asks, but neither me nor Macy is moving.

The way Sam's breathing is shortened, I can tell he's been drinking. More seconds tick by and nothing has changed. All three of us remain relatively motionless.

"Wow," Sam finally stammers, and a few seconds later his plodding footsteps retreat.

I shake my head and smile. "Sorry about that," I tell Macy.

"It's not a big deal," she replies, even though everything that has transpired throughout the evening has felt significant, part of something bigger that will offer guidance in the future.

"Go for it," Macy soon tells me.

I slow down because I don't want the moment to end. As soon as it does my head will hit the pillow and it's lights out, this evening banished to memory. More blood flow takes the decision away from me. I look into Macy's eyes, and sure enough, I go for it.

CHAPTER 34

"I got to get going," I hear while slowly regaining consciousness. Macy is putting her clothes on. Bleary-eyed, I get up to walk her to the door. We don't say much. She leaves and what happened hours earlier feels all too final. This reality hits me harder than the cool breeze blowing in off the Hudson. It won't be easy to move on from last night's events, but not doing so will only trigger the emptiness that's burning up my insides.

"Have a nice day," I tell her, then shut the door.

This closes another paragraph that won't amount to anything besides later digging deep to try to remember what her lips tasted like or the way her smile made me think twice about everything, but right now I can't deal with that.

"Good morning," I say to Sam while walking through the main room.

He says nothing.

"When do you want to leave?" I ask.

A few inaudible noises come from his mouth.

I go back in the living room and strip all the sheets from the futon, then quietly meander over to the laundry room. The extremely complicated washing machine has seventy buttons. I'm going to lose my mind if I can't start it. My eyes run over the buttons once more after putting the sheets inside.

Clueless and frazzled, I press what looks like the start button, figuring the machine is too smart to overflow and cause water damage to the entire house.

Back in the main room, Sam is quietly putting his things together, still hell-bent on not speaking. I go over to the futon and try to reassemble it, but after one lift my lumbar barks. I sigh and leave the futon as is. It will drive Sam crazy that I'm not going to leave the futon how I found it, but I'm already so fed up with his attitude that I just want to be out of this house and on the road.

"So are we—" I begin, but then am interrupted.

Sam grabs his things and heads outside without saying a word. It must be time to go.

"Did you guys have a nice night?" Daisy asks as she and her husband come to the front lawn to say goodbye.

"I did," I politely respond. "I threw my sheets in the washer, just to save you guys a step."

"Thank you. That's really sweet," Daisy says.

"Take care, guys," an exhausted Sam tells them.

"You want to leave in a few minutes?" I then ask Sam.

"I don't know. Stop asking," he warns.

By the tone of his voice, I don't dare respond. Soon Sam begins riding, and I'm meant to follow. The first hill, twenty feet from the house, is too tall so I begin walking my bike upward. When I get to the top, Sam is at the trail entrance wearing a look of frustration.

"You can't do this to me," I want to scream in his face.

Perhaps this is how he felt last night, but we are heading into New York City. Where we are going isn't a backwoods retirement community where life moves painfully slow. This is the most famous metropolis in America. We need to be in synch or else one of us might get lost.

As we ride the trail closer to the city, the noises grow louder. A beautiful day has fallen upon us, but I can't appreciate the marvelous sunshine with the vitriol gusting off Sam's cold shoulder.

The paved trail ends a few miles from the city, forcing me around a rock inexplicably placed in the center of a dirt path that leads forward.

A mile later Sam is off to the side in a park, taking pictures of a small pond. When I pull up next to him, he says nothing. He just gets on his bike as if the only reason he waited is because he has an obligation to take me to Maine.

THE LONG ROAD EAST

Then when we merge onto the first street, we are in New York City. The intensity hits me right away. Everything is fast and busy. There are no bike lanes, so we are right in the middle of traffic.

Sam misses a turn, and in order to rectify his mistake ignores the road and almost hits me while turning his bike around. I can't take it any longer.

"What the fuck?" I yell as he looks at his phone for directions.

Again, he says nothing.

We ride underneath a train, the scene a replica of the visuals from those Grand Theft Auto video games. The drivers in New York City don't accept our presence, refusing to share the road.

Everything is off. There is no chemistry between us. Sam's GPS is malfunctioning, and the blazing sun is draining my water bladder. We spend the next hour performing the same charade: ride for a bit, walk through traffic, and then get turned around somewhere in a city where strangers don't have it easy.

Where we are headed is a mystery. I only know we are meeting Sam's cousin at some point today.

Further into Manhattan, the crowds become larger, the buildings taller, and the uneasiness more prevalent. It's not a great indoctrination into New York City, but maybe there was never supposed to be. Sam gets us to a trail along the Hudson, and for the first time in ninety minutes we can move with pace.

I have never seen so many bikes, droves of cyclists scattered throughout this trail. It's like a minihighway, half of New York City riding alongside us. It shows how small the Midwest really is, because all the noise, traffic congestion, and voices of NYC collectively smash into my main sensory organ. Even while sucking down gulps of water, there's nothing to replenish the drought my emotions are trapped in. I've never been to a place like this before.

We soon stop next to an outdoor basketball court as bikes zoom past. Sam has his head buried in his GPS. Part of me wonders if his wayward navigation is a ploy designed to irritate me. Having navigation problems in a city this big doesn't make sense.

Without telling me, Sam walks through the trail and up a ramp. I don't look both ways before crossing the trail, and then a young man with dark hair squeezes his brakes to avoid ramming into me.

"Watch where you're going, you fucking idiot!" the enraged person barks.

I don't have the temerity to defend myself. This place suddenly feels ostracizing. I don't belong here. The idea of today becoming enjoyable seems impossible. I'm aware of the fact that Sam may never look me in the eyes again, but I can't let him control my emotions.

"What's going on?" I politely ask while we are parked next to a large building downtown.

"This is where my cousin works. I have to get a key to her apartment," he informs.

His cousin Amy eventually comes down to the street. She and Sam hug while I awkwardly stand off to the side.

"See you in a few hours," Amy says once the two have formulated a plan.

"We'll probably go get some lunch," Sam tells her. "Do you know any good places?"

Amy smiles. "You're in New York City. You won't have to look too far."

Sam grabs his handlebars and readies to leave.

"Where do you want to eat?" I ask him.

"I'll find somewhere."

Aligning with how my day has been going, Sam chooses what turns out to be the worst place for lunch. It's an old pizza joint, and we are the only patrons. We grab a table on the patio underneath the scorching sun so we can watch our bikes that are locked up around a small tree.

"Good afternoon," says our server, an attractive Greek man.

It doesn't take more than a few minutes to learn this guy sucks at his job. No one else is here, but it takes him ten minutes to bring out waters and take our orders.

"Can I get a Sprite?" Sam asks after ordering food.

"Sure," the server replies, but then he never brings the soda out.

THE LONG ROAD EAST

"Can I still get a Sprite?" Sam asks when fifteen minutes later the man brings the pizza out.

"You wanted a Sprite?" he asks Sam.

"Yes, please. I asked for one when we ordered the pizza."

"Oh yeah, you did. I was busy and completely forgot," the Grecian claims, even though the entire time we've been here he has been chatting up the bartender.

His ineptitude is beyond my comprehension. If it isn't the oppressing sun or the poor service, it surely is the lack of spirited conversation Sam is willing to engage in that fuels my state of irritation.

It's pulling teeth just to get him to tell me we are going to his cousin's place after eating. I have to yank a molar to learn it's another five miles away.

Five miles is nothing, but in New York City rush hour on a Friday afternoon, we won't be there anytime soon.

After paying the bill, we weave through the fracas of Manhattan traffic. A stoplight turns red, pitting us right next to Times Square. I should celebrate this fact, but none of today feels joyful. A couple weeks ago, pulling up next to the tourist attraction would have felt glorious. But today it's just a reminder that getting here was more mental fortitude than physical superiority.

Usually days on the road are fluid. One can expect to experience a full spectrum of emotions, making each day a reminder of how life is never consistent. Right now, fluidity can go fuck itself. I need a break.

"We have to go over that bridge," Sam says, pointing at the connection between Manhattan and Brooklyn.

After crossing over, the new borough isn't as upscale, but each breath also isn't filled with exhaust smoke or the distraction of a vendor trying to sell worthless items to tourists.

We come to Amy's apartment. She lives in a large unit that looks the same as every other building on the block.

"Excuse me, fellas, can I get through here?" Sam asks two guys hanging out on the steps in front of her apartment door's entryway.

"Which floor does she live on?" I ask Sam.

"Third floor," he says.

I poke my head inside the entryway and see the narrow stairway that will certainly be a pain to maneuver Joad through. Sam goes first, lifting his thirty-pound Salsa up around his shoulders, careful not to bang his bike into the white walls.

It's easier to leave Joad in the entryway and take my bike up first, so I unhook her and then meander up the stairs. The closer I get to the third floor, the more I hear loud voices coming from the apartment.

I hurry to see what's going on, walking through the door to see Sam engaged with a portly Asian man who has a stain on the pocket of his white T-shirt.

"What the fuck do you think you're doing? Why is your shit in front of my bedroom door?" the man screams at Sam.

"Look, sir. I'm Sam Johnson. I'm the cousin of Amy. I—"

"I don't give a fuck who you are," the man interrupts. "Get your shit out of my apartment. I pay rent here. This is not cool, man."

"Amy said we were staying here though," Sam counters.

"She didn't tell me," the abrasive Asian continues.

"Okay, well then I'm confused."

"You can't stay here. I pay rent here," the man wails in attempt to alleviate Sam's discombobulation.

Sam's wearing a look of uncertainty appropriate for our present situation. "Where do you want us to go?" he asks.

"I don't know. Not here," comes another vicious reply from the mouth of the irked man.

I can stand idle no longer.

"Look, sir," I chime in. "Your roommate Amy said it was okay for us to stay here for two days. It sounds like you weren't aware of this."

He suddenly looks willing to talk.

"We're not here to make your life difficult. We are just doing what we were told. I apologize for the inconvenience. We will figure this out," I tell him.

"You better," he says, and then walks back into his room and slams the door.

Right on cue, another roommate comes out with a similar look of disgust.

"How are we doing? I'm Sam, Amy's cousin."

The man does not react well to Sam's friendliness. "She only told me her cousin was coming for the weekend. She didn't say anything about bikes or another person," the other roommate reveals.

I grab my bike before more words are spoken. Staying in this apartment is no longer going to work out. Even if Amy somehow successfully pleads with her roommates, spending the entire weekend as a pariah is unappealing.

I go back downstairs and lift Joad from the floor of the entryway. There is a brown residue dripping from the bottom. It has to be rust, so I wipe it with a plastic cover. All this does is turn the floor into a murder scene.

"Fucking A," I say to myself.

Sam comes down and sees the small pool of liquid. "We need to get that wiped up," he remarks.

"I know. You got a towel or anything?"

Instead of answering, Sam takes a yellow towel and wipes up the mess himself.

"Thanks, bro," I tell him.

"No problem. I need to call Amy."

"I'll be there in twenty minutes," Amy tells Sam while we stand outside her building with our respective dicks in our hands.

The bright side to this fiasco is now Sam *has* to talk to me. He's understandably embarrassed with how everything has turned out. Yet despite everything, I sympathize with my friend. He doesn't deserve being thrust into a tough situation on account of Amy's subpar planning.

"I thought you could keep your bikes outside," Amy rationalizes when she returns home from work.

She's pointing at a small black fence right next to us. This chick must be deranged because leaving my livelihood attached to a makeshift configuration sounds like a sure way to get robbed.

Sam is equally astonished by Amy's rationale, perhaps going so far as to wonder if Amy was adopted because genetics wouldn't deal his bloodline someone as clueless as her.

"Leave them outside? In Brooklyn? Amy, come on, that's not possible," Sam reasons. "You can't be serious."

"What's the problem?" she asks. "They might be safe out here."

"They *might* be safe, Amy?" he questions, and then the two bicker.

I don't have anything nice to say to Amy, so for once saying nothing seems appropriate. To avoid being stranded in NYC, I take out my phone to begin calling Warmshowers hosts in the area.

"I'm going to the Yankees game, but you could come over after," says one man.

"What time does it end?" I ask.

"Around eleven."

"I'll let you know," I tell him.

Sam begins looking for hotels, with not much luck.

"You want my credit card number?" Sam asks one man. "Can't I just pay when we arrive? I don't feel comfortable giving you my card number over the phone."

After having that request denied, Sam hangs up and then calls a Best Western.

"It's only a few miles away," he says to Amy and I after he finishes the call.

"Good news, and we can always meet up tomorrow," Amy pipes in after saying nothing throughout our rush to find hospitality.

"Let's do that," Sam says, but hopefully he's only being cordial.

I can't spend more time with Amy. She's let us down and doesn't seem bothered. If we were somewhere where a hotel cost sixty dollars, I could move on, but this is NYC. A hotel is going to cost more than a few loose coins.

"See you later," Sam tells Amy as we ride away.

The weather has cooled when we stop at a corner store. Sam goes inside and now the only white person within sight is me. I whip out my phone to shield myself from being noticed. A few minutes later, Sam comes out with a brown paper bag filled with a six-pack, and we resume riding.

Going to the hotel is peacefully slow, a reminder of simpler times when I wasn't running around the country to mask my emotional issues. We cross under another train and then arrive at the hotel.

"Give me a minute. I have to sort this out," Sam says.

"How much?" I ask when he returns.

"Two hundred dollars."

"Fuck."

"I know, right?"

The hotel lets us shove our bikes inside a boiler room, and I let out a long sigh. It's finally time to relax. We go to the room and turn on the TV. Sam calls his mom to complain about Amy, and I scroll through Tinder to escape reality for the next few minutes.

"I can find us a Warmshowers for tomorrow night," I tell Sam when he gets off the phone. "That'll give us a chance to see the city tomorrow and then leave on Sunday."

"What about Amy? I told her we would meet up," he says.

"You actually want to do that?"

"Well, yeah. She's my cousin."

"I hate to sound like a dick, Sam, but she fucked us over today."

"I know, but it wouldn't hurt to hang out with her for a little bit."

I refrain from arguing because at least Sam is talking to me again. That alone is a win, but reuniting with his cousin sounds awful. I would rather drink rail tequila than spend another second with Amy.

CHAPTER 35

Complimentary breakfast isn't included with the two-hundred-dollar room, so we venture outside on empty stomachs. I have hooked us up with a Warmshowers for tonight, but it's all the way back in Manhattan, so we head back toward the bridge to leave Brooklyn.

The city is predictably busy for a Saturday morning. The bridge is overcrowded, making crossing over an arduous affair. As far as my eyes can see, people ignore the walk path and infringe on the bike lane. Everyone has their phones out and are desperate for the perfect shot to post on social media.

The combination of traffic and narcissism makes me want to vomit. Today was originally supposed to be filled with Amy showing us around the city, but after her egregious planning, today is just another grind.

I maneuver Joad through the street traffic, and we soon find ourselves back on the trail along the Hudson River. An old couple takes turns passing us, and then we pass them when more space opens up. After the old woman cuts me off for the third time in ten minutes, I find myself wanting to push her over so that she breaks her hip. It makes more sense now why Sam held so much animosity for the old bird back in Michigan.

"What the hell is this chick on?" I say to my myself.

My frustration lingers for a few more miles before Sam and I finally hit our exit. I can see the part of the trail that would take us out of the city. A part of me wants to leave and go hole up in a brothel with a bottle of Grey Goose. Physically, I could ride all summer, but my mental health is out of sorts. The rigors of dealing with Sam, strangers, and the suffocating heat has become too much.

THE LONG ROAD EAST

Although there were already hints of larger dysfunction, the real problems between Sam and I started back in Ithaca. My decision not to go back into Pennsylvania and continue to stay in New York burned Sam. My two dalliances also broke some of the trust we cemented.

This was the point in the trip where I didn't just talk about sleeping with women; I was actually doing it, giving Sam proper ammunition to feel marginalized and betrayed. I also sensed he was envious of my good fortune, but there was no way to prove this because he would likely never admit to those feelings.

We exit the trail and take a green bike path up a large hill until arriving at the address of the Warmshowers host. We haven't ridden more than fifteen miles, but it's so humid that already a cold shower sounds like music to my ears and satiation to my sweat-drenched breastbone.

"Let's go to that store," Sam suggests, pointing at a corner shop.

Inside the store are tiny aisles any normal-sized person would have trouble fitting through. My bony hips and big head might knock something over if I crane my neck too far one direction.

After buying some milk and two Nutty Bars, I wait outside for Sam. Halfway into the first Nutty Bar, I notice a beautiful young woman who is staring at me. Her gaze persists. She then starts walking toward me until she's three feet in front of me. Perhaps this is my lucky day.

"Are you Quentin Super?" she asks with a smile that makes me forget all about my emotional distress.

"Of course," I heartily reply.

"Hi. I'm Kristin, and this is my partner Ben."

Oh, fuck. Forget the part about this being my lucky day.

A handsome man appears from behind Kristin, and we shake hands.

"It's really nice to meet you guys," I say.

"I saw the bikes, and then I asked Ben if that might be you guys."

"Yep, it's definitely us. Sam is just inside grabbing a bite to eat."

Soon Sam walks out of the store, and after introductions we make our way to the couple's apartment building. It takes two elevator trips up thirty flights of stairs, but we finally get everything neatly settled into their apartment.

"How much do you pay for your apartment?" I ask Ben.

"Twenty-three hundred dollars a month," he says, at which point I nearly spit out my other Nutty Bar.

"That's insane, bro," I tell him.

"The only way we can afford it is because Kristin is salaried at the school she works at," Ben explains. "Other than that, I have a few big clients at my architecture firm."

Despite their vocations, these two are hippies at heart, content to throw on a vinyl record and bask in each other's presence. I look at the happiness they receive from each other and am reminded of what my hormones influenced me to give up.

The leather from the makeshift furniture in the apartment is causing me to overheat, so I begin squirming around until my skin no longer sticks to the chair. The two lovebirds are going to a concert tonight, effectively forcing Sam and I back into the company of his much-maligned cousin Amy.

"Q, if you look out the window, you can see a glimpse of Yankee Stadium," Ben says while pointing into the far-off distance.

It's then that I appreciate New York City. Not because of the faint shadow of a baseball stadium, but because there are buildings as far as the eye can see. I now feel small in a world so big. I look down from the window and notice how far thirty stories actually is.

It's a long way, just like it's been a long journey to get here. I've taken for granted the resiliency needed to come out here. There were hundreds of opportunities for Sam and I to mail it in and go back to our normal lives, countless moments we could have called it a wrap, but we never did. We still have a lot of riding to do, but reaching New York City feels like the completion of our second act.

"You guys ever ridden the subway before?" Kristin says in breaking up the afternoon melodies.

"I did in Boston," I mention, leaving out the part that when it took off on that chilly afternoon, my knees buckled and I nearly fell over while everyone looked at me like I was an idiot.

THE LONG ROAD EAST

As the afternoon ages like a fine whiskey, Sam begins making plans on the phone with Amy. I casually pretend to hear none of it. Soon, everyone is ready to leave for the night.

"Are you ready to go?" Sam asks me.

"Go where?" I say, playing dumb.

Sam has a weird look on his face. "Didn't you hear me talking to Amy on the phone?"

"No. I thought you were talking to your mom or something," I lie.

I'm petty and don't want to take the high road or any of the other obligatory gestures old people say are important. My brain is fried, and the idea of compromising with others, especially Amy, sounds exhausting.

"Well, you can't stay here," Sam says. "Kristin and Ben are going to a concert."

"I know. I'm going to come. It just would have been nice to have been asked."

"I swear I thought you heard me talking."

"Unfortunately, no, but it's fine. We can meet up with them."

No matter what, I was always going to be joining Sam. I simply wanted to have an argument about it first.

When we get down to the subway, it is delayed, so out of fear of missing their concert, Kristin takes us back to street level to find a cab. This also proves to be a hassle, so she takes out her phone and orders an Uber.

"I don't understand why there are no cabs tonight," she sighs.

A man with dreadlocks soon picks us up and quickly rips through the streets of NYC, dropping us off near Times Square.

"Just hop on the subway to get home. It will bring you right next to our apartment," Kristin assures Sam and I as she and Ben quickly scurry away.

Sam and I soon convene with Amy and her boyfriend Steve in the middle of a city park. I don't have much to say during introductions because everything I want to say violates social norms. It's better to just look around at all the things my eyes have never seen before.

The tall buildings eventually make me feel good about coming out tonight, and at one point on the stroll to dinner, while standing next to a garbage can, I grab Sam's forearm to show my appreciation for him being here.

"I'm glad we're here, man," I say.

"Yeah, it's a cool place," he unenergetically replies, his response not exactly the confidence inducer I was hoping for.

Amy and Steve take us to an Asian restaurant where she claims to have once seen Vince Vaughn.

"Everyone came up to him and wanted an autograph. I felt bad. The poor guy just wanted to get some food," she explains.

Our grand tour later features a stop at an ice cream truck. The woman running it looks like Chelsea Handler.

"You guys want to get a drink?" Amy asks in between licks of her ice cream.

"Take us somewhere cool. A place with women and a cool vibe," I tell her.

"I think I know what you mean."

I don't think Amy knows what I mean because minutes later we are sitting at an Irish bar that doesn't have more than ten people in it. The only thing Irish about the place is one of the bartenders has a noticeable accent that can be heard from our corner table because this place is *dead*.

"This is more of a happy-hour place," Steve says after everyone but me orders an alcoholic beverage.

Amy and Steve mean no harm, but they're not my type of people. They're fake and more concerned with curating the perfect images of their lives than actually living them.

"You really have to want to live in New York City if you're going to be here," Steve says in a way that makes you think he owns half the apartment complexes in Manhattan.

"What does that mean?" I ask.

"Like, you can't just live here. You need to have a job lined up."

"And you need to convince someone to rent to you," Amy interjects.

"I had to pay to fly out here to interview for jobs. Companies want to see if you have that drive," Steve says.

A lot of what Steve says makes sense, but it's the way he says it that annoys me. He reminds me of a poor person who tells people how to become rich.

"I am going to be honest," I begin. "I don't really see the appeal to living here. It's an amazing city, but to an outsider, it seems like a vanity play. Like, people live here just to say they live in New York City."

Amy and Steve look at me like I have just decoded a nuclear weapon.

"I wouldn't say that," Steve defends. "We have a lot of unique cultural things here."

"Like what?" I ask.

"Like concerts and stuff," Steve says.

"Hmm, okay. Don't get me wrong, bro. I'm not saying New York City isn't unique or a great place to be. I just think most things that happen here will also happen everywhere else."

"Where are you from?" Steve asks.

"Minnesota."

"That's small compared to here. I'd rather be a New Yorker."

"Are you even from New York?" I ask.

"Not technically."

"What does that mean?"

"I was born in South Carolina," Steve admits.

"And then your family moved here soon after?"

"No. I came to New York after college."

This guy and I will never get along.

"Steve, my whole point is, my favorite rappers will come to Minneapolis when they tour. I don't have to come to New York to see G-Eazy play."

"I guess," Steve mumbles.

Our conversation really ends there.

We pay our tabs and then walk outside, slipping past the many construction barriers that are hanging above sidewalks.

"I'm glad you said all that," Sam tells me once Amy and Steve are out of earshot.

"Said what?" I ask.

"Just reminding them that there's more to the world than New York City."

"It's sad that I have to, isn't it?"

We make a right turn. Amy begins telling Sam that she still thinks we should have locked our bikes on the fence outside her building.

"Can't you see how that's not a good idea?" Sam asks Amy.

"I think the bikes would have been fine," Steve wantonly interjects.

"Really, Steve? I don't think they would last the night," Sam replies.

"You clearly don't know New York," Steve then says.

"At least he's not pretending to," I quickly retort.

Steve rolls his eyes, and I can see Sam's frustration level is rising with each sentence that's traded.

"Don't worry. We'll leave," Amy tells Steve.

She then turns toward Sam.

"Are you going to the family reunion in August?"

"I'll be there," Sam confirms.

"Okay, well then we're going to head back. *Ciao*," she says before walking away.

"Did she really say *ciao*?" I ask Sam on the subway ride home.

"I don't want to talk about it," Sam gruffly responds.

A few stops into the ride home, a middle-aged woman with fat lips and other obvious surgical enhancements begins barking at the man seated next to her.

"Stop touching me! You're coming onto my side!" she wails.

"I'm not on your side!" the man fires back.

I'm looking at Sam, wondering when people became so sensitive. The woman then grabs her purse and scrunches into a tiny ball, doing everything she can to not have her leg rub against the man seated next to her.

THE LONG ROAD EAST

Back at the apartment, Kristin and Ben have not yet returned. I lay my head down on an air mattress and fall asleep to the noises of the city seeping in through the open window above my head. Everything today happened so quickly, yet it felt like one of the longest days of my life.

A day in New York City: eventful and draining.

CHAPTER 36

I have to get out of New York City, just like I had to get out of my shitty college town after four tumultuous years that encompassed two degrees and countless life lessons. Only this time, it's nothing against New York City. It has everything to do with me.

"How was your concert last night?" Sam asks Kristin as she brews a pot of coffee.

"It was good," Kristin politely answers.

"Tell me about it," Sam urges with a genuine smile.

I wish he hadn't said that. The morning is fast expiring, and we still have to ride forty miles.

"You almost ready to go, Sam?" I ask after Kristin tells her story.

Sam sometimes overstays his welcome, and today I'm not in the mood for his charm to engulf the entire room. The opportunity to leave NYC and get on to something new is right downstairs, but still time passes.

Packing my things into Joad is still not enough of a hint for Sam. My actions look pushy, but I'm done with words for the time being. My arsenal of diction has been temporarily compromised by an infection that's entered through a sea change of inconsistency.

"Sam, I think these guys want to get going with their Sunday," I finally muster after he asks another question warranting a lengthy response.

"Oh shit, do you guys have things to do?" Sam incredulously asks.

"I have laundry to do, but I have all day to do that," Ben replies.

"Sam, we should get going," I say, standing in the doorway like my father does when he wants out of a situation.

"I suppose it's about time we leave," Sam moans.

THE LONG ROAD EAST

We are almost out of the apartment, and then Sam's front tire pops.

"I can change it real quick," he assures Kristin and Ben.

Kristin politely smiles, but it's clear from her body language that she too is ready to move on. Sam flips his bike onto the handlebars and quickly changes his tire.

"Finally. It should be good to go now," Sam says, testing his tire against the hardwood floor by bouncing it up and down.

Sam and I are entering the third act when we step outside. During the remainder of the trip, emotions will unravel, feelings will get hurt, and the ending to this story is finally going to unfold. Tomorrow we will cross New York state boundaries, emancipating us from a territory we have spent two grueling weeks circumnavigating.

Before leaving we stop at the same corner store from yesterday.

"Fuck, that tastes so good," I say, holding a bottle of vanilla milk while Sam's thick fingers fiddle through a bag of peanuts.

It's still ungodly hot as we begin leaving the city. Sweat drips down every part of my body. I'm in the best biking shape of my life, but no matter how physically in tune my body is, it doesn't erase the pit in my stomach that churns every time I look at Sam. That fucking bastard, judging me with his outdated ideologies. If he only knew how much I love and care about his well-being, he would get over his temper tantrum and start treating me like the friend I've been the last four years.

We ride into the Bronx and take turns running over the remains of a smashed beer bottle. Pieces of the glass precipitate the paranoia that a flat tire is in my near future, but I do what I can to block out those feelings. We then get turned around on a side street in a rough neighborhood. Without telling me, Sam makes a U-turn and heads for another bridge, causing me to jackknife. My handlebars smash into the pavement, scraping my knuckles and causing a fury last felt a few days ago when Sam put his head down and shunned me the entire way into NYC.

The tension between us can't go on any longer. I would rather Sam spit in my face and deride my actions than ignore my presence. I want him to exhibit an unconditional love I swear doesn't exist, because if it did my ex wouldn't have given up on me when I put my heart before her and begged her not to stab it.

"We have to talk," I tell Sam as we stop to rest just before entering the trail we came in on.

He doesn't say anything; instead, he is peering into my blue eyes like he knows this encounter was inevitable.

"What is the problem?" I need to know.

"I…" Sam stops, examining his opening word. "It just became very apparent to me the other night that this trip is all about you."

"Which night?"

"The night in Tarrytown. You invited that girl over, and then I knew this trip was about *you* and what *you* wanted to do."

"Because I had a girl over? Why are you so against that? I wouldn't stop you if you were inviting someone over."

"Yeah, but we both know that isn't going to happen," Sam says, releasing a deep breath filled with vulnerability.

"It might," I console, but really it won't because at this stage in his life Sam has no game.

"Did you ever think what would happen if they came down and saw what you were doing?" Sam asks. "I mean, did you even for one second consider that they might not want you banging some woman in their house?"

"She's not just some woman," I defensively claim.

"Fuck. You're never going to see her again," Sam chides. "And you want that. You like slamming pussy. That's what this trip is to you. Just another chance to get your dick wet and write another book."

I pause, taken aback by his comments. They aren't inaccurate, but they don't tell the whole story. A few more seconds that feel like minutes pass before I collect my thoughts.

"I'm not going to justify what I did the other night. It was a huge risk, one I was obviously willing to take. It probably was stupid, but I wanted what happened to happen so bad."

I then pause to absorb blame that now feels all too justified.

"Look, you never even asked me how it was," I meekly say.

"What does that matter?"

"It matters because I had a damn good time, and as my best friend you are supposed to give a shit about that."

I can feel my heart sinking. The more we argue, the more I fear we cannot repair our friendship.

"This is the reality," I begin with the hope that my discourse remedies our ailments. "We can argue about this forever. I won't sit here and tell you I behaved completely responsibly because I didn't. I took a huge risk, and that was not fair to you. But goddamn, Sam. We only have a few weeks left of this, and then it's over. This trip, everything we've talked about for the last year, it's going to be over just like that."

Our eyes are deadlocked, not in a stalemate, but in a moment that can't be ignored.

"I know how we are," I continue. "We won't talk after this trip because neither of us will make the effort. And you know what? You can spend the rest of this trip justifiably pissed off at me, but what will that bring you? These are going to be the last days I likely see you, forever, and I want to spend them doing the stuff I love doing with you. I want to go back to smiling, laughing, and getting pissed that you're so far in front of me. I don't want to just *get through* these days."

I pause once again, this time to catch my breath.

"So"—and then I take a breath of fresh air—"what will it be?"

Before Sam can respond, a couple approaches.

"Excuse me," the man says hesitantly. "Are you guys traveling somewhere?"

"Can't you see what's going on here?" I want to tell him, but his intentions appear harmless.

"We're going to Portland, Maine," I say, willing to engage in some unmemorable chitchat so the guy doesn't think I'm a total jerk, even though just three seconds ago I felt guilty for wanting him to fuck off.

Soon the man and his girlfriend walk away, and I turn back to Sam. I just want my best friend back.

"It's been a hell of a trip," I tell him.

The momentum from our conversation is gone and can't be recaptured. With a slight nod of the head, Sam all but agrees that something has to change. We don't come to an agreed-upon understanding, yet still there is a feeling that the chemistry we had for so long is still salvageable.

As we resume riding, there is leftover mud on the trail from an earlier rain shower. My tires keep getting sucked into the dirt so to push Joad through I need to wake up the lumbar muscles that went dormant during the last few days cruising city streets on relatively flat terrain.

Sam is not faring well. I make it the few miles back to the pavement but turn around and see no trace of him. Nerves heighten as a few bikers pass.

One pauses and tells me, "I think your friend fell."

Unsure of what to do, I put my kickstand down and go over to the trail to glance inward. Sam is visible, but he's not moving. I can't leave my bike unattended, no matter how many times his phone continues to go unanswered. Twenty minutes pass and eventually he rolls through.

"You okay?" I ask.

"I'm fine," he says.

"What happened?"

"It's all good. Don't worry about it."

The heat and mental exhaustion make these forty miles seem endless. I don't want to be on my bike anymore. Usually during a bad ride, a few bright moments will arise out of the abyss, but I don't know if that will happen today. There is no juice or emotion to stroking the pedals, nothing that makes me want to give today my best. Instead, I'm content to mechanically move my way toward the destination.

In the late afternoon, we finally get to a city named Katonah. After ringing the doorbell at our Warmshowers a few times, we are greeted by an angry woman and two rambunctious dogs.

"Quiet! Quiet!" the woman demands of her dogs.

"Hello. I'm Quentin Super!" I awkwardly yell over the dogs' vicious barks.

"Huh?" she screams back.

"I'm Quentin Super! From Warmshowers!"

"Just come in!" I make out over the barks.

The whole scene turns uncomfortable very quickly. The dogs won't stop barking, and this woman can do absolutely nothing to deter them from howling.

"I hate when Greg does this," she complains through the deafening blows of the irksome animals. "He's never here when people show up."

I don't quite know what to say to make this woman feel more comfortable.

"We're good guys," I promise her.

She shoots me a stare that says we would be wise not to test her. The basis for her discomfort is elusive. She's both one hundred years and one hundred pounds too much for my palette, so if she has a morose picture revolving around sexual exploitation, she's horribly mistaken.

"We can just go upstairs and wait until Greg gets back, if that makes you more comfortable," Sam offers.

"Just go take showers," the woman screeches, sweat profusely dripping down her forehead.

We scurry upstairs, and I take a deep breath after escaping to the bathroom to unwind from the fracas. Within minutes, the old woman is yelling up the stairs about another problem that has just surfaced. Through the door, I can hear Sam and her go back and forth, but their words are inaudible.

"Is everything okay?" I ask Sam when I finish showering.

"Yeah. She was just yelling about towels and air conditioning," Sam explains.

Fortunately, things calm down when Greg arrives thirty-five minutes later. The dogs obey his commands, and this puts the older woman in a better mental state.

"Did you guys have a good ride today?" Greg asks.

"More or less," replies Sam.

The four of us sit down to have dinner outside, and after being mercilessly attacked by mosquitoes while feasting on pork chops, I go upstairs to relax under the air conditioning. How these next few weeks will play out is a complete mystery.

CHAPTER 37

New York state drained us. It started with getting sentimental over making it there, but in two weeks a lot happened.

I got laid a couple times.

Sam blew up over my lackluster orienteering.

He also took a mental sabbatical.

And the summer heat left me dripping sweat most nights.

We even spent a whole Friday riding through NYC, only to wind up in a Best Western.

But most importantly, Sam and I put our relationship through the wringer and came out alive.

When we finally cross the imaginary lines separating New York and Connecticut, it's a welcome relief. It's evident from our demeanors that being somewhere else on a map has sparked a reinvigoration. Happiness feels more possible than it did even twenty-four hours ago. No large cloud of uncertainty hangs over my head because the suction of the city hasn't followed my tires north. Spiritually I've been cleansed, given another chance to go out and enjoy the trip.

Today's ride is only fourteen miles, but they aren't easy. Roads still present continuous hills, and once we land in Connecticut, shoulders on the right sides of the roads disappear.

Pressure mounts on the way up one steep incline as a line of cars slowly pass. They can't all pass at once because there isn't enough room or time to completely cross into the other lane and go around. Our presence on the road is understood by all the drivers, save for one irritated idiot.

We get to the top of the hill and then he blares his horn for five seconds. I can understand his frustrations, but there is literally nothing I can do to help him besides continue to get up the hill as fast as I can.

He slams on his horn for another five seconds, and now he's gone too far.

I release my middle finger from the death grip my hands have around the handlebars and flash it his direction.

"Go around, you fuck!" I scream.

He's had a couple seconds to get on with his life, but he doesn't take it. The horn continues to disturb the early morning, and he's parallel looking at us like we are the root of all evil before speeding off as quickly and loudly as possible.

"Is that guy serious?" Sam asks.

"It doesn't matter. Fuck him. He's a dick."

Grass comes into view while homes with actual yards begin to spring up. This isn't the grass one buys and installs outside their urbanized condo either. This is the good stuff, the kind you can roll around in and let the blades poke against your skin.

It's barely lunchtime and already we are within a few miles of our next destination. I need a haircut, and we need to kill some time so that we don't arrive at the next Warmshowers too early.

There is a Cost Cutters a few hundred feet away from an outlet mall. We pull over next to the side of the building. Despite the blazing heat, Sam doesn't want to come inside.

"I'd rather just sit out here," he says.

"Are you sure?"

"Yes."

"Okay then," I say.

Sam's decision is perplexing. I thought my heartfelt speech mended fences, but his disposition suggests there are remnants of dissension leftover.

Upon entering the Cost Cutters, a blast of cool air hits my face. "Just have a seat," the receptionist instructs. "I'll call you over when we're ready for you."

Half an hour later Sam is still sitting on the sidewalk in the blasting humidity. We continue ahead to the Warmshowers spot, arriving just after 1:30 PM. Walking up to the house so early is uncomfortable. I feel like a fraud when I ring the doorbell. We didn't work hard to get here, our sweat equity not enough to have earned a free meal and a spot to stay.

A young man opens the door and introduces himself as Carl. "You guys are here a little early," he says.

"It was an easy day," I oversimplify, choosing not to tell him how difficult it can be to find a Warmshowers host at appropriate intervals.

"No problem," he says. "My parents will be home in a few hours."

Carl shows us to our sleeping quarters, which is a spacious bedroom above the garage. I throw my belongings on the bed and think about the pool outside that we passed on the way upstairs. It beckons me like many vices that send a chilling surge up and down my spine.

I then spend the next hour vacillating between the hot tub and pool. It isn't until I hear voices coming from the kitchen that I decide it's best to not completely milk this luxurious home for all its amenities.

After drying off, I walk inside and find Sam and a matriarch deeply embedded in conversation.

"How are you today?" she asks me.

"I'm feeling good, especially after taking a swim. Your pool is awesome."

"It's nice, isn't it? We had it installed a few years ago after we bought the house."

Once I change out of my swimsuit, I go and join Carl on the couch. He is watching a TV show on the supernatural.

"Are you into this kind of stuff?" he asks me.

"What stuff?"

"Spells and magic."

"Not at all," I chuckle.

"That's okay. We can watch whatever you want," he says, throwing me the remote like I have authority. "We have Amazon and all that stuff."

THE LONG ROAD EAST

After scrolling through the options, I settle on *13 Hours*, an action movie starring Jim from *The Office*. Jim is no longer the skinny guy with a bad haircut who chases after Pam. Instead, he now has bulging ab muscles and carries around a machine gun.

It feels nice to be back in Netflix mode and away from the labors of riding. Feelings of shame and deception continue to bother me because we didn't ride very far, but at the same time, today is an anomaly. Tomorrow we will be right back at it with a lengthy ride.

A few hours later, the entire family sits down for dinner on the backyard patio. I like the vibe but without just cause fear the dad hates me.

"Have you guys enjoyed your trip so far?" he asks.

"Yeah. It's been hard, but today was an easy day," I say.

"Where did you come from?"

"Katonah."

"That's not very far," he says, his nose wrinkling.

"Yeah. Like I said, it was an easy day."

To avoid further interrogation, I bury my feelings in the food until my stomach can expand no further.

"You guys have to be out early tomorrow," the dad then says.

His tone is harsh.

"That's fine," I assure him.

"Like very early. I leave for work at 7:00 AM," he says with a straight face.

The dad isn't the first person to be demanding. Out east, Warmshowers hosts feel more like landlords than they do friends. This is because I've heard enough stories about other cyclists who abused the good-natured Warmshowers system. The actions of those before us had rightfully angered these generous hosts, creating a carryover effect that now forces them to be stern with their expectations.

"LeBron is such a pussy," Carl says later that night as the Golden State Warriors begin their final assault of the Cleveland Cavaliers in the NBA Finals.

"What do you mean?" I ask him.

"LeBron always complains. He's so bad."

I should remind Carl that LeBron James is the greatest basketball player of all time, a human being that is 260 pounds of complete athleticism. This in no way makes LeBron a "bad" athlete.

Yet somehow I feel these words would be lost on a kid who plays lacrosse.

CHAPTER 38

Something is still bothering Sam. He turned off at a fork in the road and has disappeared, once again leaving me without an idea of how to proceed.

"Where the hell did he go?" I whisper to myself while standing on the shoulder.

There aren't any obvious clues as to where Sam has gone, so I decide to stay on the main drag. Ten minutes later, a pickup truck is humming down the roadway too close to my left leg, its tires overlapping the white line.

"What the fuck!" I yell at the vehicle.

The truck then tries to completely cut me off, its front right tire further encroaching the shoulder and bolting in front of me. My hands pull on the brakes, careful not to get sucked up into the back tires.

The truck beams ahead and comes to a dead stop in the shoulder. As I roll forward, it dawns on me that I might have to fight this guy.

Pulling up just behind the rear, the driver gets out with a smile on his face. "Hey, sorry about that," he says. "I didn't know how else to get your attention."

I hold my tongue and think of a million other ways he could have gotten my attention that didn't involve trying to run me into the tall grass of rural Connecticut.

"I just talked to your friend back there. He thinks you might have missed the turn," the man then says.

"My friend? The chubby guy with a big beard?" I ask.

"Yeah, that's him," the man says.

My eyes roll, unable to comprehend how Sam could be so careless yet again. "I told him I'd try to find you and bring you back," the man continues.

"How far is it to get back to where he is?" I ask.

"A few miles."

Backtracking is painful, and I have been going downhill since the fork in the road. The idea of riding uphill back to meet Sam sounds worse than picking at a pimple on my forehead.

Looking at my phone, no service bars pop up.

"Who do you have?" the man asks.

"T-Mobile."

"Yeah. That probably won't work until you get to Bridgeport."

"I figured."

Mine and Sam's trip is closer to the end than the beginning, yet we are back to making the same mistakes we made in Wisconsin.

"What do you want to do?" the newly revealed volunteer firefighter asks.

"Could you tell him to meet me at the turn up ahead? Tell him it is a really fast couple miles."

"Not a problem," he says.

I shake the man's hand and scurry ahead to the turn the man says will intersect with the backroad Sam wants to take into Bridgeport. Putting trust into a stranger isn't the ideal choice to make, but my gut that won't sag until middle age urges me to believe his words.

Now at the turn, and with time to kill, into my stomach goes a bag of Pop-Tarts, followed by another. Fifteen minutes then pass, and still Sam has not shown.

From where I stand, the turn is clear as day. There is tall grass behind me and a house across the street. Apart from that, nothing will confuse Sam regarding where to turn.

More time elapses and perturbation creeps in. If I'm going to die out here, it will be from riding in the wrong direction; not because someone's vision goes astray and rams their pickup into my ribs while I'm standing on the side of this quiet road with a package of strawberry Pop-Tarts in my hand.

THE LONG ROAD EAST

With this thought in mind, I forge onward until hitting a small town. It's so dead it only needs some dust and a saloon to look like a place that rednecks frequent. There is still no cell service, so I wait a few more minutes before once again feeling that dying in the middle of Connecticut isn't a cool enough story to write on my tombstone.

A sign at the intersection gives no indication which direction Bridgeport is, so I turn right because that runs parallel with the county highway. The more I travel, the further away from Sam I potentially become. Right now, I'm just looking for a bar to pop up on my phone so I can dial Sam and coordinate an exit strategy from this self-induced predicament.

Further along, the road continues to feel as if it is taking me in the opposite direction of where I need to go. It's then that I hit a construction zone.

"Do you need directions?" the police officer overseeing the site asks me.

"Sort of, sir. Is there any way I could borrow your phone?"

"Sorry. I don't get service out here," he says.

"Shit. Well, do you know how I could get to Bridgeport?" I ask.

"You have to go back and take a right at the stoplights. That will bring you right into the city."

"I appreciate that."

"But be careful," the cop warns. "You don't want to get caught going down the wrong street in Bridgeport."

I follow the officer's directions until arriving just outside the city. Heeding the man's advice, I stop because if someone sticks a gun in my face, I'm going to vomit and start bawling.

Fortunately, after taking my phone out of my pocket and switching off airplane mode, a few bars have cropped up.

"Where are you?" I ask Sam after getting a hold of him.

"Bridgeport," he says.

"Of course, but which part?"

"I'm by a mechanic shop. Not gonna lie, it's sketchy."

"Send me your address. I'll get there as soon as I can," I say.

Quickly I receive a text, and after plugging in the coordinates, the map says Sam is on the other side of the city. It's hot as hell, and asinine blunders like these make me want to have a mental breakdown.

Under the scorching afternoon sun, sweat starts dropping down my balls, back, and the nape of my neck. My janky phone just has to make it a few more miles and then my insecurities concerning directions can be put back into Sam's Verizon-fueled navigation.

The cop wasn't lying. As I resume riding, the neighborhoods get worse. Not a single person is walking on a sidewalk or watering their flowers. It's desolation, the kind that turns a suburban white boy's confidence into that of a five-year-old girl.

The miles slowly tick down, and my phone is on life support. At stops I begin memorizing the street names in case my phone dies.

With half a mile to go and no more turns, I flip airplane mode back on, and soon Sam comes into view. He is indeed stationed right next to a rundown repair shop that is encased by barbed wire fence.

Despite the ominous tones blanketing the area, I can't help but feel a sense of overwhelming relief.

"You're right. This place is super sketchy," I tell Sam, feeling like we haven't seen each other in years.

"I know. I've been standing here for twenty minutes sweating my dick off."

"Well then let's get going, bro," I cheer.

The gas station we stop at for lunch doesn't sell food, just candy and pretzels that couldn't satisfy even a type 2 diabetic. To combat this misfortune, I buy only fluids: iced tea, a slushy, chocolate milk, and then a few waters. Predictably, these items do nothing to satiate my hunger.

"You're not going to believe this," Sam says while we camp outside the gas station. "It's so hot my phone shut off."

"Jesus Christ," I tell him.

I am running out of answers for today. Beads of perspiration drip down the small hairs on my groin and delicately drop into my bike shorts. This evokes nostalgia from when my mom yelled at me every summer to "go outside and sit in the sun. You got to get those zits off your back!"

It's *that* hot outside.

Sam's chest hair is sticking out from his blue shirt, the perspiration from his upper body reflecting the sunlight directly into my eyes.

Just outside New Haven, Connecticut, we stop alongside the ocean. Our Warmshowers host is still at work so we need to wait for 5:00 PM to arrive.

"How do you see women in your life right now?" the newly magnanimous Sam asks as we sit on a bench overlooking the waves splashing against the shoreline of large rocks.

"Woah, man," I huff, caught off guard by Sam's desire for introspection. "I mean, you know how I've been since my relationship ended. I think the main thing is I don't want to be in another relationship for a while, mainly because I don't want to let anyone down."

"When do you think you'll start looking for another girlfriend though?"

"That's a good question, Sam," I answer, looking off into the ocean for an answer. "Right now, I love being single, but I do miss certain parts of being in a relationship."

I stop to think a bit.

"I guess that's why my relationship ultimately failed, you know? I couldn't pick a lifestyle," I state.

"Interesting answer," says Sam. "That's a decision you'll have to make soon enough though, right?"

"We'll see," I say.

As the two of us reflect, it's easy to begin thinking about the end of our time together. I have taken for granted how easily this trip could have never happened. Sam and I are just two dudes from Minnesota and South Dakota, the crossing of our paths never guaranteed. Had I gotten into my preferred university, I wouldn't be here. Or if Sam continued to build bridges and not return to school, that too would have derailed this moment.

It's a gift that we're here, even if we don't spend all our time laughing and opening presents.

As I catch his eyes gazing out at the ocean, it's nice to appreciate how life is unfolding. Life isn't going smoothly or making my wildest dreams come true, but it's happening. I'm not merely a passenger to a narrative anymore.

"It's about that time," I tell Sam when I check my phone.

"Agreed," he says, standing up and revealing patches of sweat that have been smothered all over the wooden bench.

We finish the ride into New Haven and arrive at the host's apartment as rush hour traffic descends upon the city.

I'm snacking on a bag of gummi worms when a foreign finger taps me on the shoulder.

"What's up, guys? I'm Lee," a new voice says.

I turn around and see our host, a ginger who is dressed neatly in a suit after having just finished his daily dabble in the 9–5 world. "Bring your bikes inside," he urges.

I disassemble Joad in pristine order and bring her up into the apartment.

"You guys want to see the sights before we eat?" Lee asks while we congregate in his kitchen.

My stomach is rumbling, but I agree to go sightseeing anyway. Lee then goes to the fridge and opens up a twelve-pack.

"You guys like beer?" he asks.

"I'll definitely have a beer with you," Sam inserts with excitement.

After the two touch cans and chug a highly-caloric IPA, we walk outside to Lee's car.

"I do have to admit, Sam, it really is cool being able to take apart Joad so easily," I mention on the way to the car. "I'm glad you were always on my ass about that kind of stuff."

"See! That's what I had been trying to tell you the whole time!" Sam recalls of the last six-plus months, my acknowledgment long overdue. "Doesn't it feel nice to master your equipment?"

We get in Lee's car, and he takes us all over New Haven, including up a mountain that overlooks the entire city. It's a pleasant view, but like Mount Rushmore, after two minutes it's time to move on with life.

"You guys got a Chipotle here?" I ask when we get back to the car and return to New Haven proper.

"Fuck Chipotle," Sam chirps from the backseat.

THE LONG ROAD EAST

"What's wrong with Chipotle, Sam?" Lee asks.

"Everything. We can't go there," Sam bristles.

"Sam's on a diet," I sarcastically add.

"No, I'm not," Sam doesn't fail to mention.

"I was going to say…," Lee says hesitantly.

"We don't have to get into it. I just don't like Chipotle," says Sam.

"How 'bout pizza? You guys like pizza?" Lee asks.

"I'm fine with that," Sam says.

We settle on a nice pizzeria not too far from the apartment. The three of us dump exorbitant amounts of carbs into our stomachs, quickly devouring both pizzas that were brought to the table.

"I got this," Lee announces when the bill comes.

"Definitely not," I say upon realizing the total is close to one hundred dollars.

"Look, I know where you guys are at. I've been on the road before," he insists.

"That's too much though," Sam chimes in.

Sam and I need to be cognizant of money. We want to go to Vermont, but right now the budget is gradually shrinking. Not letting Lee pay the bill will stretch our finances out, but it's also not right to let him cover the entire check.

"Okay, we can split it then," Lee offers.

"That's fair," I tell him.

Back at the apartment, the three of us begin to develop more camaraderie.

"What are you going to call Sam in your next book?" Lee asks me as the beer begins to distance him from sobriety.

"I haven't thought about it, man," I say, sweat streaming down my forehead while we roast in the living room.

"Don't listen to this idiot," Sam tells Lee. "All he's going to talk about is me cockblocking him and how I'm the biggest douche in the world."

"Sam, don't play games. You know your cockblocking skills are unparalleled," I jab.

"Fuck off."

"Guys, listen. I have a great name for Sam," Lee says while standing up.

"What's that?"

"Chester Dunston."

I rollick on the futon, the dubious look on Sam's face enough to make my stomach ache with laughter.

"I'm going to have to write my own book," Sam responds.

"Write it after Q's comes out," Lee suggests.

"Just so I can clarify all the stupid shit that Q is going to write," Sam finishes.

"He's just mad about his fling in Michigan," I say while holding back tears of joy.

Sam and Lee both look at me, one puzzled, the other ready to explode.

"What happened in Michigan?" Lee asks.

"Not much," I say. "Sam made a bunch of suburban dads die of laughter."

"What's so crazy about that?"

"Well, there was this guy named Ruxin—" I say before Sam cuts me off.

"Don't be bringing up that idiot," Sam snaps.

"No, I want to hear this," Lee says. "What's the deal with this Ruxin guy?"

"Okay, Lee. Here's the thing. After the little bonfire party we were at, Ruxin drove us all home."

"What's so weird about that?"

"That alone, nothing. But see the thing is, Lee, I'm not so sure our friend Ruxin actually went home. Sam was staying in a school bus and—"

"Okay, no! Shut the fuck up! That's not what happened!" Sam intercedes. "Lee, this guy is an idiot. There was this fucking creepy guy. And yes, his name was Ruxin, but—"

"Ruxin might have went into the bus, but no one besides Sam really knows," I counter.

"Jesus christ!" Sam howls. "No. That guy is a fucking weirdo and he left. End of story."

"Geez, bro. Don't get so defensive," I tell Sam, who by now is heavily under the influence of a locally crafted libation.

"You guys crack me up," Lee says after finally catching his breath. "But I'm going to bed. Catch you guys in the morning."

All night, the laughter has shielded us from the reality that it's too hot to sleep, but as I now snuggle into a grainy futon, the suffocating humidity has me guzzling any water I can get my hands on.

I turn over and a litany of empty beer cans sit on the table next to the TV. I don't understand how those two shoveled all that beer into their stomachs, but when I see Sam's belly rise and fall with each breath as he scrolls through his phone, there is no doubt where all the liquid has been deposited.

CHAPTER 39

"I just kept drinking until I fell asleep. It was too damn hot in there," Sam says the next morning. "And how shitty was it that Lee had AC in his room," he continues, his head shaking as the receptors in his brain search for sobriety. "And we were stuck on the floor."

"Yeah, yeah. I get it, bro," I say between chugs of chocolate milk. "I know it was hot. The problem was you drank too much beer. You're dehydrated."

"Fuck off, Q. You don't have to remind me."

Later that afternoon, just before we get to a man named Jeff's house in the coastal town of Westbrook, I hit a wall. The idea of sharing pleasantries is too much. Being berated by Sam sounds better than formal introductions with our new host, but cordiality still has to be conducted.

"Good to meet you," I tell Jeff after putting down the kickstand on my bike.

Jeff is hairier than Sam and more energetic than a child ramped up on Mountain Dew. After dropping our stuff off inside, he takes us to the local beach with his girlfriend Ming.

She's the antithesis of Jeff, in that she's antisocial and afraid of the sun. Her and I sit in silence on the sandy beach while gazing at our respective partners parade around in the ocean like a bunch of walruses.

Sam and Jeff can't stop smiling, even as the hard waves collide with their pasty white skin, impacts that send my untouched skin into horror. I notice Jeff has these funky-looking nipples, perhaps a byproduct of age, or maybe a bad stroke of genetic luck. They look ready to bite Sam if he turns his back for too long.

THE LONG ROAD EAST

Fifteen minutes into their childlike fun, my ears start to ring, pulsating like a speaker at a rock concert. My feet haven't taken one step in the water, yet my eardrums beg for shelter and ibuprofen.

"You get earaches from the water?" Jeff asks after he comes out of the water soaking wet and hovering too close for comfort.

"I think it might be the saltwater," I explain. "In Minnesota we only have freshwater."

"Yeah, maybe! I once knew a guy...," he says, but my ears drown him out, both because they're in pain and also because I can't stand to hear another story that took place decades ago.

My ears are still trembling after we leave the beach. Once we're in the car, my dick suddenly becomes hard. Not the good kind of hard either. Seeing Sam and Jeff play peek-a-boo in the sea was not arousing. No, this is the type of annoying erection that comes on when it's cold and windy.

As I lean forward in my seat to conceal my embarrassment, we drive to an ice cream shop across the road. Sugar would seem to be the perfect midafternoon slump buster.

"I really like ice cream," Jeff awkwardly tells me as we all stand in line.

"Me too," I say, scanning the menu of overpriced desserts.

"I really want the vanilla fudge," he intimates.

It sounds like he wants me to buy him a cone.

"I'll have the cookies and cream," I tell the person behind the window.

Jeff leans in closer, but he's not going to get what he wants. I should buy him a cone but opt not to because I don't have any money to spare. The woman hands me my cone and then I walk to an open space.

"You guys want to get back home?" Jeff asks.

"That's probably a good idea," I say.

The car bumps up and down the same raggedy road Sam and I came in on. I go upstairs and try to drain my untimely erection, carefully trying to maneuver my dick so that it doesn't scrape against the bowl of the toilet.

"Fucking A," I tell myself, wishing that I wasn't super tall so that I could just stand and urinate like a normal man.

For dinner, Jeff cooks some frozen fish and green vegetables that came from the freezer downstairs. It tastes awful, and my stomach won't be satisfied unless American beef ends up on my plate in the next five minutes.

"How's the food?" Jeff asks.

"It's good," I lie, but not even Jeff buys my attempt at social nicety.

Not much more is said, until near the end of dinner Jeff begins to tell a story.

"I biked across Israel," he says, going on to explain his time navigating a war zone.

The entire time his girlfriend Ming sits awkwardly at one end of the table, not interested in actively listening or joining the conversation. The entire tenor of this meal is off-putting.

"How about a movie?" I ask after Jeff finishes his story, looking to my right and motioning toward the TV sitting in the living room.

Any movie, no matter its quality, would be a relief. Anything to save me from the monotony of forced conversations.

"There are a few DVDs under the TV. Go see if there's any you would like to watch," Jeff instructs.

There are only three movie selections lying on top of a stack of magazines, the most appealing being *To Kill a Mockingbird*.

"How about this one?" I say, holding up the DVD.

Everyone agrees, and soon Jeff pops the movie in the DVD player, but even this becomes a drag because he is narrating the entire movie. I begin to fantasize about what it would be like to run away from this house.

"Beautiful movie," Jeff says when the film mercifully ends.

"Something like that," I smile, then head downstairs to finally get out of my own head.

CHAPTER 40

"You guys are more than welcome to stay another night," Jeff says the following morning after our bags are already packed.

"Sorry, Jeff, but I do think we have to get going," I tell him.

The disappointment in his eyes is surprising. Nothing about our time together warrants a second invitation. Maybe Jeff wants Sam to take his shirt off again, but even that feels like a stretch, considering the deepest our conversation got the previous night was talking about the weather in Minnesota.

"You sure?" Jeff asks one more time.

"I'm sure, but thank you again for everything."

"You know, Q. You and Sam remind me of some people I had stay over a few years back."

"Really? Who were they?"

"It was a group of seven girls," Jeff says, and immediately I'm lost.

"I'm struggling to see the connection," I admit.

"The thing is, Q, those girls and I, we had such a good time, they ended up staying the entire weekend!"

Jeff tells this story with so much joy you could be forgiven for thinking that all seven girls took turns giving him a blow job.

"That's awesome," I say, then conveniently look back at the house and see Sam strolling out.

"Have a good one," Sam tells Jeff as we begin riding away.

"You too! Call me if you need anything," Jeff yells as him and Ming watch us pedal away.

Our fling with shortened rides continues. We go only another thirty miles to Groton, home of the much-renowned Paul's Pasta. It's there we meet a teacher named Beth who has just finished her last day of the school year.

"Excited to go back next year?" I ask her.

"Not at all. I'm actually quitting. Gonna go work at a bank," she smiles.

"Why are you going to go work at a bank?" Sam asks.

"I just really don't want to be a teacher anymore. The kids are always sick, and then as a result I get sick. I'm over it," Beth informs.

"That makes sense," I say.

"But enough talk about me, guys. Come on, let me show you the house. My husband will be home in a few hours."

After unwinding, we all gather in the kitchen. "Do you have any recommendations for dinner? We heard Paul's Pasta was good," Sam says.

"It's *the* place to go in Groton," Beth claims.

"Then that's where we're going," Sam replies, and soon we're out the door.

When people go to a pasta restaurant, they typically have an idea of what's on the menu. Tonight, that's not the case for Sam. He's practically having a heart attack while perusing the present dining options.

"What would you go with?" Sam asks the cute, sandy blonde-haired waitress. "I've never been here before."

"There are many choices," she says, her impatience evident as her eyes peek around the busy dinner crowd.

There is so much uncertainty in Sam's eyes it looks like he will shed a small tear if he orders the wrong thing. I want to nudge his ribs and tell him to hurry it up, that I'm hungry and this waitress isn't the human version of Google Reviews.

Sam does this often, getting so worked up over making the perfect choice that he slows the whole restaurant down. It doesn't matter if we're in a five-star restaurant, or a Pizza Hut at the local strip mall, Sam will still find a way to yap a server's ear off while debating the array of available food items.

Eventually Sam decides on the chicken carbonara. The pastas come out a short while later, and we eat without much fanfare.

"That was an awesome meal!" Sam exclaims when the check comes.

"Eh, it was okay."

"Okay? No, Q. That stuff was bomb."

"Sam, it was a bowl of pasta. Calm down."

"I don't get it, Q. You're a spaz when it comes video games and women, but then when it comes to food, you don't give two shits about what it tastes like."

I laugh. "Dude, it's just food."

"I'll never understand you," Sam remarks with frustration, and then we leave the restaurant.

Like two idiots in a maze, after getting back to the house, Sam and I can't figure out how to open the side door. No one is home, and despite playing back Beth's instructions in my head, they make no sense. The key she gave us doesn't even fit in the lock.

"This makes absolutely no sense," I tell Sam. "Can you see where I'm going wrong?"

"Not really."

Sam then takes his turn with the key, to no avail.

"I really have no idea either," he says, handing the key back to me.

It then hits me that the house has two doors.

"Why don't we just go to the front?" I ask, and for one of the few times all trip, my intuition is correct.

Later that night, while walking to a brewery, Sam and I have the necessary talk regarding how we're going to finish the trip.

"If you want to go to Vermont, we can't go out tonight, because I am *bleeding* funds," Sam says.

"It's that bad?" I ask.

"Yeah, unfortunately."

"I do want to see Vermont," I admit. "But it's only for the vanity of saying I went there."

"I get it. Personally, I don't care if we go to Vermont or stay along the coast. I'm just going to leave it up to you," Sam says.

"Shit. Putting me on the spot here."

This decision is a tough one. If we skip going west back through the mountains, we will be done in less than a week. Ironically, there are still four more states we have to hit, but even that can't delay the inevitable baggage check at the airport.

After a few blocks of mulling it over, there is still no clear answer. Part of me thinks we should go to Vermont while we are out here. Maybe that valley chick I dream about awaits in a wild forest. But that's all a fantasy. Reality is less glamorous. One scroll through my checking account reminds me that it's time to go back home.

"It's probably in our best financial interests to go home," I confess before we hit the brewery, my head drooping.

Sam gives me a pat on the back. "Take your time, dude. You're probably right about leaving, but hey, we can always hop on a plane and come back out to Vermont. Maybe even this summer," he encourages, even though we both know that won't happen. "It's totally up to you, but like I said, we have to go back to the house right now if we're going to Vermont."

Sam is giving me all the power, and now that I have it, what to do with it is perplexing. We stand on the top of a hill and look down at an ugly brown strip mall where the brewery sits. There is no alternative to this quandary. I have to make a decision and then stick to it.

"We can skip Vermont," I tell Sam. "It'll give me another reason to come back out here."

A weight has been lifted, but I don't feel whole.

"Did you guys check out that brewery?" Beth asks when we return.

"Yeah, but it was busy. Nowhere to sit," I reply.

"I'm sorry to hear that. You guys want to go out for a drink with me and Victor after he gets out of the shower?" she asks.

"That would be great," Sam answers.

"Let's do it. I'm already hungry again," I tell them.

Over dinner, Victor begins to share his own biking story. "I made it from coast to coast in six weeks," he unenthusiastically shares in between sips of beer. The more he describes his journey, the more mechanical his trip sounds.

"Doesn't sound like you enjoyed it very much, Victor," I say.

"I did. It was just a lot of work."

"At least you had the wind on your back though."

"Right? Otherwise I would have given up," he says, then looks happily at his wife and gives her a small hug.

Not even the ice cream that comes out after my burger and fries can stop the depressing thoughts reverberating inside my head. The end of this trip is near, and I can't help but debate whether the right decision was made by forgoing Vermont.

CHAPTER 41

The morning sun is bleeding in from the window and smacking against my forehead, offering the perfect lounging spot while I relax on the couch and scroll through my phone.

As Google Maps goes through an update, the front door opens. It's Beth.

"I forgot something on my way to work," she laughs. "I finally get to go clean out my desk."

"Right on. I hope you have a good day," I reply.

Beth grabs her forgotten item from the counter and then stops. "Q, do you have a girlfriend?" she asks.

"I don't. Why do you ask?"

"I have this coworker of mine. She's been looking for a guy."

"I'm sure she's great, Beth, but I am going back to Minnesota in less than a week," I sigh disappointedly.

"What's your point?" Beth asks.

My head snaps her direction, and I sit up straight. "My point? Well, like, I couldn't, you know, *date* your friend."

"Who said anything about dating? Just go have fun."

I continue to look at Beth, unsure if this is a practical joke or if I really am this lucky.

Seeing my facial expression, Beth laughs. "Don't worry. She's fun. I'm sure you two will have a great time. I'm going to write her number down and leave it here on the counter."

"Cool. I'll pick it up before we leave," I nonchalantly say, but as soon as Beth shuts the door I bolt from the couch toward the tattered piece of paper.

Today, Sam and I are back to riding in the sixty-mile range. Fatigue sets in on the hills, and all day the pouring rain continues to suck the life out of my soul. With only a couple miles to go, the Friday rush hour traffic is heavy, adding to the onslaught of rain that continues to smack against every patch of my exposed skin.

At the final left turn, cars blaze past on both sides, gusts of wind smacking against my face. A warm breeze then breaks through the cold front, awakening my senses and altogether offering a reminder that sitting on this chunk of pavement is a privilege. Arriving at this juncture was never guaranteed.

For many years I caved to others who didn't have my best intentions in mind. It's not to say they were oppressive or malicious, but no one got me on a bike and told me to bike across the country. Many urged not to, but if I listened to those people, I wouldn't be here. It's not quite that time of the journey to spill my heart out, but it's time to write the speech.

Right now, I don't want to meet any new people. Either the world has become less friendly, or I have turned into a hermit who prefers the comforts of solitude. Isolation is now only possible on my bike, a vessel that grants me freedom from the outside world. Even though I continue to operate within the same hemisphere as everyone else, my universe exists through the prism of a man-made machine that transports me from state to state.

It's lonely out here, strapped to a bike and enduring seven weeks of experiences that not many will ever be able to relate to. But this is the path I asked for when I daydreamed about my own success.

Still, as this trip winds down, my past glamorization of a highly applauded future is not coming to fruition. Fame hasn't followed me around the country. In a few short days, I will return to my normal life as the guy who biked across half of America.

"Are you guys doing okay?" a woman in a gray sedan asks as we rest under gray clouds on the side of the road.

"We're fine," I promise. "Just trying to battle through the day."

We eventually stumble upon a yellow sign in front of a driveway that says we are on a bike path. Intuition tells me it is our avenue out of this long and murky day full of rain and self-imposed destitution.

"I think we're here," Sam says, pointing at a small house on the right.

A couple dogs come bounding through the hissing rain to say hello as we pull into the driveway. Our host Ralph also comes out onto his front porch and waves us forward.

"Take your clothes off before coming inside," he tells us after we shake hands, which grants me another visual of Sam's caveman-like back.

Ralph is old and his wife is out of town, leaving us three all alone on this dreary, overcast evening. Many characters have passed before my retinas lately, but few are as laid-back as Ralph. We don't have to say much to be able to spend time together in peace.

After dinner we gather around the fireplace. "Can I have the Wi-Fi password?" I ask Ralph while wrapping a warm blanket around my body.

"No problem. I just need to retrieve the password first," he answers. "You okay?" he asks after returning from his office with a small slip of paper containing the Wi-Fi information. "You look like you've seen a ghost."

"I'm fine. Just a cold day," I try to assure him.

Nothing is physically wrong with me. It's all emotional, my brain continuously circling around a woman I think I still love. It's hard to convince myself I still have feelings for her, that despite me being here and having engaged in wildly promiscuous behavior, my heart still belongs to her.

It may sound silly, but I have love to give. You can't clearly see it, but then again, neither can I.

CHAPTER 42

While relaxing in bed, I begin texting the mystery number that Beth left for me. After a few exchanges, we agree to meet in Providence.

"I'll text you when I get there," I tell her.

"Fantastic! See you later," she replies.

For breakfast, Ralph fries an egg and then blends it like a smoothie before drizzling the remains onto a piece of bread.

"Is this like an East Coast thing?" I ask him.

"I don't think so. I've been eating this my whole life," he responds.

He hands me a plate, and the protein-packed concoction tastes surprisingly good, providing the energy my starved muscles crave.

Today has a vibe. Not only am I going on a blind date, Sam and I will also be staying in a hotel, courtesy of a beyond generous Warmshowers host named James. A few days earlier he offered to put us up in a hotel because the spare room in his house was being used by an old friend.

"You totally don't have to do that," I had written at the time.

"I want to," he wrote back. "Just keep the room under two hundred dollars."

"That certainly won't be an issue."

"And how about we meet for lunch? Just so I can put a face to your name."

"Let's do it. We're going to be in Providence in a few days. I'll give you a call when we're in the city."

After leaving Ralph's home and then beginning our ride, clouds still hover above us. Fortunately, the trip into Providence proper goes fast.

Sam is a couple hundred yards ahead speeding down a hill when he cuts in front of a car that's creeped into the crosswalk. The car slams its horn, and Sam has a less-than-enthused gaze for the overzealous jerk operating the steering wheel.

When it is my turn to pass the intrusive car, I swerve to avoid an altercation, in the process bumping over a curb and crushing my left testicle on my seat. That's nature's indirect way of telling me not to be a coward.

After taking a trail for the next few miles, we pull into the hectic city and find a heavy dose of Saturday-morning traffic awaiting. This presents a good chance to pull over and call James.

"Are you guys by Brown University?" he asks.

"I have no idea, but I can use my GPS to get there. We can't be too far," I tell him.

"Sounds good. Get to Brown, and we can go from there."

At no point in our conversation did James mention that Brown University, much like Cornell, requires one to bike up a mountain to reach the school's perch. I again lose Sam on this northbound jaunt that reinforces one's place in the social hierarchy, so shackled by Joad that I contemplate cutting her loose and watching her roll back down the hill to join the rest of the peasants at the bottom.

I get to the top and receive a text from James. He says to meet him at a restaurant that isn't far from the school. After catching up with Sam, we head to the restaurant, arriving before James.

"We should pay for this lunch," I tell Sam while we lock our bikes against another tree.

"That's a good idea," he agrees.

Once we get inside the restaurant, it becomes clear that we're not at the local Applebee's. Judging by the class of people in here, along with the highly decorative wallpaper, this place carries an aura of decadence and sophistication. I look down at my yellow jacket that's seen better days and hope we don't get told to leave.

"Two?" the hostess asks me.

"Actually three," I say, and then she leads us to a table tucked into the shadowy corner of the restaurant, lest we been seen by the more fanciful patrons.

THE LONG ROAD EAST

"Okay, so no matter what, we have to pay for this meal," I half-heartedly repeat to Sam once the hostess lays the menus down and walks away, dying a little on the inside while picturing at least fifty dollars limping from my wallet.

"Don't worry. We'll get through this," Sam nervously chuckles.

James soon arrives in a green convertible. I wave at him and motion to our table. When he gets closer, his mouth looks a little wonky.

"I had a dental procedure done earlier this week," he explains before anyone can comment on the weird shape of his lips.

James quickly establishes himself as a likable guy who is interested in our stories, and this takes the early tension out of the room.

For as calm and in control as James is, it's hard not to pity him when the topic of his divorce is broached. While money may not be of consequence to him, that seems to matter little because he talks about his divorce with the same emotion as if it just happened yesterday. It's sad to see another person added to the long list of men who have fallen victim to ex-wives searching for a different kind of love.

I'm not as jaded, yet.

"And here's your drinks," our attractive server says. "The food will be out shortly."

"I don't mean to be crass, but our server is so sexy," I acknowledge when she walks away.

This woman is not just beautiful. She's the most beautiful woman I have ever seen. I'm not talking girl next door, I'm from Minnesota, and most of our women could be linebackers for the Vikings either. This woman belongs in Hollywood, making millions of dollars and snorting lines of cocaine in the bathroom before she goes on Jimmy Fallon.

She walks with a purpose, the buttons on her shirt covering up something memorable, yet leaving enough skin for the imagination to run wild. Her white pants accentuate every part of her feminine essence. She has two hoop earrings and just enough eyeliner to perfectly match her hair. And it's all crystallized with a subtle East Coast accent. From a vanity standpoint, she's the whole package.

"It's hard not to be captivated by her beauty," James starts. "That's how she chooses to present herself."

His response is surprising. It's as if he could care less that this woman is attractive because vanity plays no role in his decision-making process.

"You don't care that she's hot?" I ask him.

"Q, I'm forty-six years old."

"I don't get it."

He leans back in his chair and then adjusts his glasses. "You guys are too young to understand this, but there is so much more to women than just their appearance."

I feel slightly offended.

"I don't say this as a criticism. What are you guys? Twenty-seven, twenty-eight?" James asks.

"Twenty-five," Sam says.

"Okay. Yeah, so you guys are young. You're supposed to be chasing women right now. But one day you will look back on this part of your life and laugh, and you'll have a clearer view of what's really important."

It's then that the server returns. "Here's the smoked salmon," she says, then puts a plate down in front of Sam.

We eat our food and continue exploring other aspects of our respective lives before an hour that feels like only a couple minutes passes.

"Are we thinking about dessert today, guys?" the waitress asks while clearing away the plates.

My eyes are unable to break from hers.

"I want dessert," James states.

Dessert sounds marvelous, but the prospect of paying twelve dollars for a measly slice of cheesecake isn't practical.

"We maybe want to be careful," I tell James. "Sam and I want to pay for this meal as—"

"You're not paying for this meal," James asserts.

"We should pay though. You're getting us a hotel."

"No, don't worry about it. I come to Minneapolis from time to time. Next time I'm there, you can take me to dinner," James says.

I look at Sam, but he doesn't know what to say either. "Okay, if you insist," I say. "Also, I've been looking for hotels. The cheapest I can find is sixty dollars. Is that okay?"

THE LONG ROAD EAST

"Don't worry about price, Quentin. Just get two rooms that are comfortable for the both of you," James then says.

"That's really nice and all, but I don't want to take advantage of you."

"You're not. Just book the hotel."

"But the other day you said to keep it under two hundred dollars," I remind him.

"So? What did I just say now?" James says sternly.

I'm taken aback, momentarily unable to respond.

"I suppose if you're okay with this," I mutter, then scrolling through my phone and finding two rooms at the Best Western that totals over four hundred dollars.

"I'll meet you guys at the hotel," James says after everyone finishes their last bite of dessert.

Sam and I go outside and unlock our bikes. James speeds past in his convertible, and I wave goodbye.

"Am I missing something here?" I ask Sam.

"I think the guy just has some money to blow and wants to give back. It doesn't seem weird to me."

"Seriously? The dude has just dropped like five hundred dollars on us."

"Sure, it's a bit much, but truthfully I believe he really does just want to help out."

We ride a short distance to the Best Western and are then greeted by James and a few employees.

"I've been telling the manager here about your guys' trip," James says, then pointing at the man in charge.

"He has, and I want you guys to come down to the restaurant for dinner tonight," the manager says. "It's on us."

"Wow. That's really nice. You don't have to do that though," I say.

"It's no problem. Your friend here told us your story, and we think it's awesome, so please, come on down tonight," the manager reiterates before then walking away.

The lottery has drawn my number today. Free lunch. Free dinner. Free hotel room. I can't ask for much more.

"I'm going to take off then. You guys don't need anything else?" James asks, not sticking around long enough to let Sam and I annoyingly keep thanking him.

"I think we're good," I say, finally willing to just let things be.

James is a class act in every sense of the term. Fortunately, we kept in touch after that day. I sent him a copy of *The Long Road North*, and later received a postcard of him on his bike in Times Square.

"Your book inspired me to do a trip of my own," that postcard read.

I never thought there was anything materially I could give that man, but I was wrong. We all have something to give.

***

We put our bikes in another storage room and then go our separate ways before dinner. Sam is only one room over, but within a few minutes I can't help but feel lonely.

"When do you want to eat?" I text him.

"You must miss me," he replies.

"No. I'm just curious," I lie.

"Sureeeeee."

I want to be near Sam, if only so I can annoy him with my intentions for the evening. It's difficult to process this empty feeling, but I'll have to get used to it because in a few short days Sam will be gone.

My dear friend Sam could have given up on me all the way back in the beginning of our trip. He could have seen my outdated bike and slow pace as a reason to jump ship and do his own thing. But he stuck with me, and not even because he had to. We would have remained friends even if he took off and left me to fend for myself, but if there is one thing in the world I can never have too much of, it's good people.

I owe Sam a lot, especially as it pertains to getting us all the way out east. He's fixed my bike on more than a few occasions and directed us out of trouble numerous other times, but that hasn't been his biggest impact.

Sam's genius has been his ability to pick me up when I'm at my lowest points. Many times I wanted to slow down, take a break, or delay the inevitable pain that would come from climbing another mountain or bucking through gruesome gusts of wind, but Sam never let me get to the point of falling and being unable to get back up.

I'm lucky. Simple as that.

After a few hours, Sam obliges my pleas for company and sustenance, and we go downstairs to eat at the hotel restaurant. I order a spicy mac 'n' cheese that tastes worse than tuna casserole. If this food wasn't free, coming here would have been a waste of time.

"You going to behave tonight?" a female bartender asks a man a few seats over who is drinking a bottle of Coors Light.

"That's why I'm drinking this," he laughs, jiggling his bottle.

"Okay. I'm just making sure," she cautiously tells him. "I don't want a repeat of the other night."

Sam then nudges me in the shoulder. "Dude, check this chick out!" he exclaims, showing me a woman on Tinder whose name is Burpee.

"I guess you're not asexual after all," I tease.

Upon closer inspection, I determine that Burpee isn't real because a) no one with a brain would name their child after an exercise routine, and b) all of Burpee's photos are grainy and look like they were pulled from Google images.

"She can't be real," I tell Sam.

"I'm pretty sure she is," Sam responds.

I want to hug Sam and tell him everything will be okay, that he's beautiful in his own special way because for some reason he has this uncanny ability to match with fake profiles on Tinder. Worse, it takes him until they ask for a credit card number before it finally clicks that the women aren't legit.

"Damn!" Sam predictably wails minutes later.

"Fake profile?" I ask.

"Yeah! Those fuckers!"

I throw my right hand up, roll my eyes, and try not to snort a laugh. "Just keep plugging away, bro," I encourage, even though I am willing to mortgage my hairline that Sam is not getting laid tonight.

It's not that Sam is weird or ugly, although sometimes he wears undershirts, and I've never seen a guy pull an attractive bird when a white tee sticks out from the collar of his top garment.

Instead, Sam's issue is rather simple. When it comes to sealing the deal, he hasn't yet developed that killer mentality needed to accomplish these types of ventures.

To clarify, the killer mentality is that "it" factor, that part of your personality that comes alive when adversity hits. We all showcase it in different ways. Geriatrics exhibit theirs when they go to Walgreens and the coupon for 5 percent off skim milk won't scan properly. In that case, the old-timers don't leave until they've literally been spared seventy-three cents. This "it" factor is what satiates our vices, but it can also be used in productive ways.

Sam unleashes his when he rides his bike, but when he gets around women, he cages up and replaces it with this gay-best-friend act I can't bear to watch.

Everyone has that friend, the guy who isn't direct enough and becomes way too nice around women. This guy thinks he's developed a new way to get laid, and that us mere mortals that fight for superiority on the douchebag food chain are behind the times. But in reality, he's already lost and is going nowhere near a woman's vagina.

Nothing makes a woman feel less horny than when you laugh at all her jokes and look at her like you're just in it for the conversation. It's too bad this method fails. You would think being nice and smiling at appropriate intervals would be the best way to get women, but in my experience, the more direct and self-serving one is, the more interested women become.

It's worth noting that if you want a girlfriend, being nice will probably get you there. And by *there*, I mean the Chicago Cubs eventually got there (there being a World Series ring), but it took them over a century to do it. Don't hate women and don't hate the player. Sadly, the game makes no sense.

THE LONG ROAD EAST

One of the other bartenders at the restaurant, a balding man in his late thirties, tries to impress two middle-aged men by pouring a Grey Goose bottle upside down into a rocks glass. During his routine, he winds up spilling an ounce of the hallowed vodka on the floor, and my heart breaks.

"I gotta go," I tell Sam.

"You don't want a Grey Goose before you go?" he asks.

"Nah, I'm good. Most of it is on the floor anyway. Plus, that girl I'm seeing tonight will be here in an hour."

"Don't have too much fun," he jokes as I leave him to his glass of beer.

I don't know much about the woman I'm about to go out with, other than that her name is Lisa and she's a teacher. I have only seen one photo of her, but in my world that grants her the same amount of validity as a full background and DNA check.

She soon arrives at the hotel, and when I go outside, fortunately she is even more attractive than her photo suggested.

"Where should we go?" she asks as we speed through the highway back into downtown Providence.

I shrug my shoulders. "To be honest, I have absolutely no idea. I'm not from around here."

"We'll just drive then," she says.

In a few minutes we come to the area where Sam and I earlier met James.

"Let's go to that place," I say, pointing at a new restaurant across the street with beer signs on it.

Lisa parks her car and then we walk inside. I set my elbows on the bar, and they become wet thanks to the leftover condensation from the previous customers.

"How's it going tonight, guys?" asks an approaching bartender with a nice beard.

"Very good. Can I please get a Long Island without tequila?" I ask while wiping my elbows on my clothes.

Lisa orders a beer and then we talk about the basic, compulsory topics that are typically broached on most first dates. She likes her job, and even more ironically, was offered employment while at a bar with her boss.

"That's different," I say, having never heard of a teacher becoming gainfully employed over cocktails.

"What do you want to do now?" she asks as our drinks slowly become empty.

"I want to hold your hand and walk around for a little bit," I admit.

"I'd like that," she says with a smile.

We then walk around downtown, taking in the residential neighborhoods and beauty of the night before stumbling upon a large group of rambunctious people still left over from a gay pride parade earlier in the day.

"Did you go to the parade?" I ask Lisa.

"No. It's not my scene," she says.

I smile at her, at the same time acknowledging a connection is taking place. Holding hands with Lisa doesn't feel like a chore or part of a routine. It's a powerful injection of lust, foreplay before the foreplay that my hormones tell me is inevitable.

We walk back up a hill toward Brown University, stopping at a bench to admire the campus and relax.

"How did you see this night going?" I ask, wrapping my arm around Lisa's shoulders.

"Well, I kind of thought we would get some drinks and just get to know each other," she smiles.

"Really?" I chuckle, not buying her plausibly deniable sentiment.

"Yeah. I figured we would just go out and have a good time."

More time passes and Lisa burrows her face into my chest. It's completely endearing, and when she looks back into my eyes, I grab the back of her head and kiss her.

"Do you want to go back to my hotel?" I then ask.

"Yes, I do."

I'm fortunate to be in this position, especially because pre-East Coast I spent too much time sulking over my inability to get laid. If I had the foresight to know what was coming, I might have relaxed on the swiping and self-induced internal bleeding. But going through all those empty nights was necessary to realize that things in life have a way of working themselves out.

Once Lisa and I enter my room, I take off my shoes and pretend to attend to a small order of business in the corner by the desk. This gives Lisa a chance to settle in and feel comfortable. I turn around, and she is standing near the bed closest to the door. I walk over and put my hand on the small of her back and begin kissing her.

What a privilege it is to be alive.

CHAPTER 43

"Are you hungry?" I ask Lisa while looking out the window of my hotel room.

"A little bit," she says.

I look across the street and see the golden arches. "Will you eat McDonald's?" I ask, scrunching my face and hoping the question isn't construed as offensive.

"It's not my first choice, but I guess that's okay."

"Sounds good. I'll be right back then," I tell her.

I quickly leave the room and head over to McDonald's, the idea that Lisa may leave while I'm gone not lost on me. The restaurant inconveniently resides across a busy street. The stoplight is taking forever to turn green, but eventually I am able to cross and make it through the doors.

"What can I get for you?" a tired-looking young woman asks when I reach the counter.

"You guys don't have a dollar menu?" I answer back, my ordering skills proving to be rusty after a long hiatus from fast food.

"What is that?" the woman asks.

I look at her strangely. Everyone and their mother know what the dollar menu is.

"Usually you can pick items off that menu, and they only cost one dollar," I tell her.

She looks at me like I'm crazy, and her patience is thinning faster than my dad's hair pre-Y2K.

"Do you know what you want, sir?" she asks again, adjusting her glasses to accentuate her impatience.

In a hurry to get back to Lisa, and the fate of the dollar menu not altogether important, I order four sandwiches, some hash browns, an orange juice, and a two-dollar chocolate Frappe. The total is twenty-three dollars.

"Is that number correct?" I ask the woman, unaware that it's even possible for one person to spend that much money at McDonald's.

"Yes. Are you paying by cash or card?"

Once again I look at her, stupefied. "Card," I say, then move to the back and wait for my food.

"Did you know it's possible to spend twenty-three dollars at McDonald's?" I ask Lisa when I return to the hotel, my amazement on full display while taking out food from the bag.

She politely laughs. I'm talking while my mouth is full because bursts of joy keep streaming into my brain telling me life is great and it might never get better than this.

"Eating out is expensive in Rhode Island," she notes.

"Evidently. Remind me never to come back here," I joke, but based on Lisa's perplexed expression, she doesn't understand my sense of humor.

After the second McMuffin, my stomach goes into a frenzy. In an hour I'll have to lock myself inside a bathroom. Silence then engulfs the room, but the greasy sandwiches and large orange juice has revitalized my energy levels.

"What are you thinking about?" Lisa asks when she sees the musing expression on my face.

"I think we should have sex one more time," I say.

"I'm okay with that," she grins.

When it's over we lie on the bed. The third time has sapped all the spirit from my legs, but Lisa moves out from under the covers and begins putting her clothes on.

"I got to get going," she says, and her words elicit a tinge of disappointment.

"For sure. Do you have a long drive home?" I ask.

"Not really, but I have things to do today."

We walk to Lisa's car and then I hug her. She drives off and, just like that, becomes a memory. Part of me wants to go back upstairs and reflect on what just happened, but there isn't enough time to let my mind shoot down five different rabbit holes of self-examination.

So I just choose one wave of thoughts to battle through, and on the walk inside it becomes even more obvious that this trip is going to end, and my debut book is going to hit the shelves. Already feeling vulnerable, now a portion of my life is going to be on full display for the world to see. What people read and then say will be a direct commentary on the way my life has thus far been lived.

Being in this position is what I signed up for: to be ridiculed, and if I'm lucky, appreciated. Of course, I only envisioned the positives when I signed that publishing contract and dropped it in the mailbox on a cold winter night last year. That should have been a celebration, a small stack of papers being shipped to New York City beginning my career into the unknown. But something was missing then; namely, someone to tell me I was doing the right thing.

Fast-forward a few months to now and something is still amiss. Biking out here in the pursuit of something bigger hasn't made me happier. It's changed me, and I can't say for the better.

Sam isn't answering his phone, and after the third time, I pack all my bags and head downstairs to sit in a storage room with a young man who tells me he has three kids from three different women.

"I used to party too much, man," he explains.

I don't know how to respond because many of the decisions he made I don't agree with, but at the same time many of the things I have done also warrant reprehension.

Just a few months prior I seemingly had it all together. A woman who loved me and a writing career in development. I threw away half of that equation to be single again.

Sadness runs through my eyes while reflecting on how I got here. So much utter failure on account of my actions. It's not fear of judgment that hurts the most. It's knowing that I deserve to be right where I am.

When Sam finally comes down to the storage room, he's in rough shape.

"You go out last night?" I ask.

"Yeah. I went to that pride parade thingy downtown."

"Well, that explains that guy walking into your hotel room last night," I quip.

THE LONG ROAD EAST

The young man next to me laughs.

"Yeah, right," Sam groans. "Did you bang that chick?"

"Duh, bro. Don't ask dumb questions."

"I figured," Sam mumbles, and then we exit the back door of the storage room to begin a thirty-mile ride into Massachusetts.

We don't get far before Sam pulls over at a Dunkin' Donuts, and when he comes out of the restaurant, he is carrying a gigantic chocolate drink that he soon slams with fervor.

"You going to make it there, Sammy?" I ask.

His eyes squint, and he lets out a deep sigh. "You know what, Junior? Just focus on the ride today."

"I want to make sure you don't crash again though."

"Q, I'm fine. I just need a little boost before we keep going."

When we resume, I pull ahead. The further I go, the more I keep looking back for Sam out of fear that something bad might happen to him. He's had multiple bad mornings these last several weeks, but few that have lingered as long as the one on display today.

After craning my neck ten times, I finally see Sam laboring closer. His approach is more sloth than Seabiscuit, and his face is deep in his handlebars. I want to launch into a diatribe laced with frustration and disgust, but my poor friend is suffering.

"Are you okay?" I ask when he gets closer, but he blows past without saying a word.

A strenuous day on the bikes is our comeuppance for cheating the road these past days. We haven't packed on the miles to the same extent we once did, and now we are getting lazy in our preparation. We have too much confidence in our ability to get from city to city, and this has impacted our preparation.

I had sex twice this morning because I can ride forty miles in my sleep. Sam pounded extra beers last night for the same exact reason. We should be going into Portland on a tear, proving to ourselves why all this was even possible. Riding like amateur frat boys on a bender of epicureanism is not paying homage to those who stood by us when we set out to do this trip.

A couple hours later we arrive at the next Warmshowers site. New hosts Carrie and Steve present the opportunity to put all our camping gear to use for the first time.

"There is a nice patch of grass for you to set up your tents," Carrie says, showing us to a small open area next to their home in the woods.

"This is going to suck," I tell Sam when Carrie walks away, looking at the dried grass stained with dog piss.

"Get your stuff out, Q. You're sleeping outside tonight," Sam reminds me.

Needing my tent, I reach into Joad and brush away sand and dirt that accrued on the road. Joad is in rough shape, on her last legs by the way the silver on her frame is rusting. Her plastic flap that was supposed to keep rain off my gear is tattered and in need of replacement. The flat tire she endured back in Erie, Pennsylvania, didn't add years to her life expectancy either.

Assembling my tent, which weeks ago was effortless, is now alien. Patience is hard to come by, and after a few feeble attempts my puppy eyes come out, and I look toward Sam for guidance.

"If you need help, just ask," he scolds.

"A little help would be nice," I hate to admit.

Despite all the instability sleeping outside will bring, it beats sleeping indoors with our hosts who thus far have not created a friendly environment. We've already been asked to stay outside on the patio with an army of buzzing mosquitoes, and at this rate, it's only a matter of time before I dial up the nearest Best Western.

Amicability is hard to muster as the four of us later sit on the patio and pretend we care about each other. I should grow up and at least fake an interest in their kids who are running around, but it's hard to do so. It's much easier to ruminate in my chair and let my body language do the talking.

"Is there anywhere I can charge my phone?" I eventually ask both Carrie and Steve.

"There is a spot right behind you," Steve says, pointing to an outlet barely visible along a brick wall draped with weeds.

"Thanks."

THE LONG ROAD EAST

I then call American Airlines to see if they will accept my bike as a second checked bag.

"Yeah, we can do that," the representative with a soothing voice says. "When you're at the airport, just tell them it's a checked bag."

"Thank you so much. That is really nice to know," I tell the man.

A few more details are sorted, and now this trip has an end date. A countdown would be appropriate if not for the lump that forms in my throat as a consequence of this finality. My dream of beginning a new life out east isn't going to happen. Eighteen months of planning and subsequent action will soon conclude. It's down to a few precious days with Sam, moments that will unmercifully strip me of the exhilaration these seven weeks have provided.

In a way, going home is a relief. Physically, I began mailing it in after Vermont was called off, and emotionally, I'm fried. There are only so many people to meet and places to see before the reality of a dwindling checking account is simply too much to ignore.

In a desperate attempt to not have the night be completely weird, Carrie and Steve take us to the beach. Parking spots are rare, and the beach is more depressing than a gloomy day in Kansas. The sand on the beach is clumpy, and the only women wearing bikinis should be wearing one-pieces.

I don't want to be here, but the alternative is sitting in my tent brooding and getting eaten alive by mosquitoes.

"Do you guys like ice cream?" Carrie asks us once the beach affair has faded.

"Definitely," Sam says.

"Okay, good. You'll love this place by our house," Carrie guarantees.

She's talking about a famous ice cream farm, but I've never heard the words *farm* and *ice cream* in the same sentence.

We drive to the place, and it is indeed a farm, save for the small modern building that looks like a food truck. People packed ten rows deep are waiting in line.

"This place must be pretty good then," I say to Carrie.

"It's awesome," she claims, but twenty minutes later, the line has barely budged.

"This must be *really* good ice cream," Sam says.

"I don't know why the line is so slow," says Carrie.

"Is there a Dairy Queen around here?" I ask.

"Hold on, Q. You don't want to wait?" Sam asks.

"Not really. To be honest, I'm tired."

"There's a Whole Foods nearby," Carrie offers.

People coming from the front of the line brush against my shoulder. They have smiles on their faces and ice cream in their hands.

"I'm sorry. I don't know what the problem is. I'm just impatient," I tell them both, my tolerance thinner than Nicole Richie circa 2006.

"I'll get Steve and the kids. We can go somewhere else," Carrie says.

Sam looks one more time at the lengthy lines, then agrees. "Yeah, this is too long," he admits.

After venturing to the grocery store to load up on sugar my body doesn't need, we head back to the house for dinner. Steve is grilling a steak, and the sky makes it look like rain is in the forecast for later this evening.

"You guys like steak?" Steve asks us.

"Of course," Sam says.

Steve takes the meat into the house and comes back out with two plates.

"Thank you for cooking us steaks," I tell him.

"No problem. Sorry we only have the one to share," he replies.

Steve goes back to the grill and turns the propane off. He then plops a small piece of steak on my plate. I look at him like he's crazy. One petite sirloin is not going to feed Sam, myself, and his entire family.

"Anyone want a beer?" Steve then asks before heading inside.

"I'll have a beer with you," Sam replies.

He comes back out with a few beers, Carrie lingering behind him as a warning to her husband that this evening better not kick it up a notch.

"You're sure you don't want a beer?" Steve asks me as he struggles to pry the top off the bottle in his hand.

"I don't like beer, Steve, but thank you," I tell him.

Twenty seconds later, Steve still can't get the cap off the beer. He looks so out of place in his struggle, like he hasn't had any friends come over in the last few years.

"Do you need help?" I ask.

"No. Just give me a minute," he grunts.

After some more painful attempts, Steve finally gets the cap off the beer bottle. "That one was really on there," he smiles.

"Hey! There we go," Sam cheers, downing half the bottle with his first swig.

"Can I get you anything else? What about root beer?" Steve asks me.

"I'll do a root beer. Thank you."

The three of us don't talk about much. This evening once again feels like a charade, that we're just here to appease Steve and Carrie's desire to give back to the world. I soon trudge off to my tent and climb in.

After forty-five minutes I already can tell this will be a long night. I'm hungry, cold, and my back is about to encumber a litmus test of grass-related horrors. And then rumbles of thunder can be heard crackling from the sky above.

Fuck this.

CHAPTER 44

"My back is so goddamn sore!" I howl the next morning.

"Really? I slept great," Sam claims, his satisfaction an annoyance to my cringing lumbar.

Today we are going to spend a few hours in Boston before riding to a suburb with our next Warmshowers host. A couple months ago I flew out here to spend time with a woman I met at a desolate bar, but after the good times faded, she too ended up deriding me and my cavalier philosophies.

Coming to Boston now is neither a romp down memory lane nor a portal to originality. We're just here, floating through existence.

Sam and I kill a few hours at a small restaurant next to a park before riding over to the Boston Harbor. We stop next to a 7-Eleven that is flooded with throngs of tourists.

"I got to grab something to drink. It's too damn hot out here," I tell Sam, and then walk inside.

When I come out of the store with a Big Gulp full of blue Gatorade, I see Sam talking to a young couple. I learn they're travelers without a destination, just two lost souls trying to figure life out.

"Where are you guys coming from?" the man asks Sam.

"St. Cloud, Minnesota," he responds.

"That's cool. I hope you guys find what you're looking for," the man says.

"Good luck to you as well," Sam tells them, and then they walk off.

Sam and I walk over right next to the Boston Harbor and find an empty bench underneath the shade. We still have some time to kill before our host Chip gets off work, so we start talking about our plans for once the trip is over.

"I hope Teddy isn't as spastic as he was in St. Cloud," I say to Sam, referring to a motorcycle lover I will move in with once these last few state lines have been crossed.

"What do you mean?" Sam asks.

"I just hope Teddy is cool with me bringing women over, and if we have issues, that we can talk them out and move on."

Sam grimaces while peering into the harbor.

"The first time we have any issues I'm going to sit him down and make him work things out," I proclaim.

"That seems unrealistic," Sam sighs, inhaling before adding more words. "That's the thing, Q. You're always doing things your way. You're so intent on doing you, it makes you a real tough person to be around."

"What do you mean?" I ask.

"Well, a good example is if you don't want to do something, you say so right away, and then that just kills the mood. Like in Buffalo, you totally shot down Chuck's Skee-Ball idea, and then you asked him if you could bang some Tinder chick in his house."

"I was being honest. That was kind of the idea, Sam."

"Whatever, man," Sam intimates, but this discussion isn't finished.

Peace between us is maintained, but no treaty has been signed. There is still a struggle for power. Sam's description of my personality intersects with my own self-assessment: toxic, abrasive, tough to be around, a pain in the ass. I should change, but being twenty-five doesn't yield itself to outside opinion.

In the heart of Boston's rush hour, we later ride to the corner of a busy street. Our host Chip arrives wearing glasses and a helmet.

"What do you do for work, Chip?" I ask.

"You guys ever heard of Draft Kings?" he asks.

"For sure. That's a cool place."

"Well, I work in the same building as that company."

"Oh," I mumble, disappointed he doesn't actually work for Draft Kings.

"You guys have anything I should know about for this ride?" Chip asks.

"I have this trailer," I say, pointing at Joad. "So don't try to weave between cars, otherwise I'll get stuck."

"Noted. Okay, fellas, my wife has dinner for us back at the house. Let's get going."

The three of us leave the city via bike paths that are littered with cyclists whose only concern is getting home from work.

"I will say"—Chip starts as we wait for a light to turn green—"it's impressive you guys have all that weight and are speeding past people."

"We've been carrying all our stuff for the last six weeks. The road has certainly got us in shape," I tell Chip.

"Either way, you guys can ride."

"Thank you very much," I say, a small smile forming after receiving such a nice compliment.

A few turns later the brisk ride is over.

"You guys are going to have to sleep outside tonight," Chip deadpans as we walk to the door.

"No problem," I tell him, but internally frustration writhes.

"Normally you could stay inside, but my son is here, so there is no room for you," he explains.

I can only assume the son must take up half the house, and that's why we're relegated to sleeping outside. Even a carpeted floor would beat mosquitoes, dewy blades of grass, and using a sweatshirt as a pillow.

"Honey, the bikers are here," Chip yells up the stairs to his wife.

"Okay. Give me a second," she yells back down.

Everything is normal until I go upstairs to shower. Faint footsteps of the wife hovering around the bathroom can be heard. I get the impression she is uncomfortable with my presence in her home.

"This is so awkward," I whisper while looking at myself in the mirror.

As I peel my clothes from my sticky skin, I can now hear the wife rummaging through a closet just outside the door.

"I'm not going to steal anything, especially not your hair straightener," I think about yelling.

Even as the warm water of the shower connects with my face, she's still outside the door. Something is awry.

When I finish and head downstairs, Chip and Sam are on the patio drinking beers. I sit down and then Chip's son Matt arrives. Matt is much like the Rex dude from back in Michigan, in that he's a huge douchebag with zero interest in politeness.

Matt shakes Sam's hand and casts him the gaze reserved for an ingrate, then affords me the same pleasantry. I want to punch this guy in the face.

"Can I go shower, Chip?" Sam wants to know.

"Actually, you have to wait. My son gets priority," Chip replies.

Then the wife comes out fully loaded. "So how far did you guys ride today?" she asks, not making eye contact with anyone.

"Thirty miles," Sam shares.

She looks at Chip like they have been duped.

"We've kind of been taking it easy since New York City. We've been on the road for almost seven weeks now," I jump in.

"Oh, I see," the wife says, her dissatisfaction with our answers very apparent each time she wrinkles her nose and then casts Chip a concerned gaze.

She then walks right back inside, almost as if the only reason she came out was to pass the time. I look at Sam to gauge if he also thinks the situation has become unnecessarily hostile, and he too is wearing a look of concern.

"Everything okay?" Chip asks as he takes a gulp of his beer.

"Everything is fine," I try to assure him.

Soon chicken wings and corn are served for dinner. The wife and Matt come out to grab some food, then immediately retreat back inside.

"You were saying it's been a tough trip," Chip says as he deposits a thigh into his stomach.

"Yes. Lot of miles and a lot of fighting," I chuckle, and then the wife comes to the screen door.

"Chip, can you come inside for a moment?" she asks.

"Of course. One second, dear," Chip says. "Excuse me, guys. I'll be right back."

I turn to Sam. "This is weird, right?"

"We might have to get a hotel," he acknowledges.

A few minutes later Chip returns like nothing has happened.

"Chip, we aren't being an imposition, are we?" I ask.

"No, no, no," he assures, going back into his chicken without addressing his reason for leaving.

Not knowing what to do, I simply shrug my shoulders, and then we finish our conversation about the rigors of a cross-country trip.

Later that evening, a light sprinkle begins to fall as Sam and I relax on the patio. Sitting outside becomes humiliating as the rain quickly turns into a downpour. Even a spot on the kitchen floor would trump cozying up on the small patch of grass in their backyard.

Small beads of rain creep through the leaves as I turn to Sam. "Can I speak freely?" I ask him.

"About what?"

"There are just some things I need to get off my chest," I reveal, diving back into the conversation from the harbor.

Sam positions in his chair the same way a guidance counselor would. "Sure, Q. Go ahead."

"First of all, I know you still have things you want to say about that night in Tarrytown, and I promise you will get your chance, but I need the floor for the next couple minutes."

Sam leans back in his chair, raindrops splattering on his shoulders, and takes a deep breath.

"Okay, so I want it to be known that I wasn't cheating you on effort in the early stages of our trip. I was behind because my bike sucks. And I know this because our legs were moving at the exact same speed," I say.

"The whole time I'm wondering why you're going faster," I continue. "I'm working just as hard, if not harder. I know I'm a better athlete than you, so nothing was making sense. But I don't want you to think I cheated you on effort."

Anger begins to form on Sam's face. Backlash will soon come.

"And then what happened in Tarrytown was amazing. I just want that to be known. I had a great time, and I am pissed that you ruined that moment. I am pissed that you let that affect the ride the next day.

And I'm pissed that you didn't try to talk to me about it. Plus, I still don't understand why we hung out with your cousin after she clearly fucked up. I know she didn't mean to, but she put us in a bind."

Time to switch topics.

"And I don't know if I've gotten so much better as this trip has progressed or if you've just regressed," I clamor, in reference to Sam's dwindling status as the front man for our rides. "But I don't know how I'm sticking with you now."

I take a deep breath and prepare for the storm that is now coming my way.

"Are you done?" Sam asks, his face looking like Darth Maul.

It hits me then that I never thought about Sam's response when I formulated these thoughts in my head.

"You can go," I say a bit nervously.

Sam rolls his eyes and shakes his head before speaking.

"Wow. So clearly there are some things you've been holding back. I'm surprised, to be honest," he begins, his thick hands not yet gripped around my neck.

I want to go back and preface everything with the obligatory "this is not personal" sentiment, but it's too late for that.

"You remember that first day we left St. Cloud?" Sam asks. "We stopped at that gas station, and you were saying you were already sore."

I nod.

"Q, I knew then that this trip was going to be all about you. You had no weight, I was fully loaded, and we were still neck and neck. And then the next day, we can't even get out of Maple Grove, a place you grew up, without getting lost," Sam says, his decibel level rising with each word that stabs through the irritating raindrops.

"And then we're on that trail and we got passed by four women. Four women, Q! Fuck me! You were already struggling, and I realized this trip was going to be a nightmare. And then you're walking your bike up hills near Minneapolis. Like, what the hell, man? Don't get off your bike in that situation," he lectures.

Sam then switches onto the most personal of topics.

"As far as regressing on this trip, wow. I don't think I've regressed, but you clearly think differently."

"That's not to say you're not riding well," I safely remark.

"Hold the fuck on. It's my turn."

A blitzkrieg of rage is inevitable. "Look," Sam says. "I knew you'd catch up at some point. You'd figure out your system and you wouldn't lag behind, but then you started chirping me going up some of those hills, and that shit pissed me off."

"Dude, that was a joke," I defend.

"Really?"

"Duh. Do you actually think I thought I was riding better? I'd beat you up hills and then you'd fly right past me on the way down. It's just the weight difference."

Sam veers another direction.

"But that thing in Tarrytown. That was all about you. You weren't thinking of me. You weren't thinking of what would happen if they found out. You just wanted to get your dick wet the whole time. That's all this trip has been about. A chance for you, like I said in New York, to chase pussy and write another book."

"Dude, you have to get over that. I didn't come all this way just to write another book," I fire back. "If that ends up happening, great, but I did this because I like riding my bike. And I want to be a writer, so another book is part of what I want to do with my life. But in no way did I sign up for this trip just so I could write a book. We planned this long before I even wrote the first one."

"Whatever. This trip is still all about you."

"How though, Sam?"

"Everything is about you, Q! The pace, eating every ten minutes. All this," he says, throwing his hands up and looking around. "All this is so you're comfortable."

"There is no way you can honestly say that. I've made sure you're included in decisions. And you dictate how we get places, so it's not all about me."

"Q, it started at my cousin's place. I mean, I'm just hanging out enjoying time with my family, and then suddenly, there is this chick at my cousin's house that you're trying to fuck."

"You know her! And your cousin said I could invite her!" I say as my anger intensifies.

"It doesn't matter. I knew you were going to do what you wanted, and I was just along for the ride."

"No, dude. Fuck that. That's not how this trip has been."

"Okay, so not taking the trail through Wisconsin, that wasn't your call?"

"I made that decision for the betterment of us. You were going to be fine regardless, but I would have gone crazy on that trail."

"I wanted to go through those tunnels. And to have to tell my cousin we weren't because you didn't want to was embarrassing," Sam says.

"It's not even their trip. Why does it fucking matter?"

Sam's eyes pop, and he is about to unleash a haymaker.

Suddenly, Chip opens the door.

"Fellas, can you, uh, keep it down?" he asks. "We have to work in the morning."

"We're just talking," I say.

"I know, but just talk a little quieter. My wife is getting annoyed."

Sam and I look back at each other but don't say another word. Another fight cut short before the final bell.

CHAPTER 45

"Good morning," a chipper Sam tells me the next morning. His friendliness is a mini miracle, or else it's part of a manipulative ploy I can't yet figure out.

"Morning," I respond skeptically. "You going to be ready to go soon, dude?"

Sam is shuffling through his bags. The clouds from last night have vanished, leaving only blue skies and the burning sun.

"Damn," he muses.

"What's up? Your tire flat again?"

"No, it's not that. I can't find my camera," he says, tossing all the items in his bags onto the lawn.

"Sam, I don't want to be that guy, but did you look everywhere?"

"Yes, of course."

Sam then gets up and goes to knock on the back door.

"Oh, you're still here?" Chip says upon answering the door.

"Yeah. Sorry to bug you. I just can't seem to find my camera. Mind if I look inside real quick?" Sam asks.

"I suppose. I guarantee it's not in here though," Chip assures.

Sam disappears through the entryway, leaving me to deal with Chip for a few more minutes.

"Thanks again for letting us stay," I say as he stands at the door.

"It's not a problem," he says back, looking inside as if time is of the essence.

A few minutes later, Sam comes back empty-handed.

"If you could send me your guys' addresses and contact numbers, that would be great," Chip says.

"Sounds good," I tell him, altogether having no intention of actually sending any information to this man.

THE LONG ROAD EAST

"I am not going to send that guy my information. Fuck him and his dumbass family," Sam laughs while we brisk down a hill and away from calamity.

"That was the worst night of the whole trip," I declare.

"I swear they took my camera. I had it right there on the counter by the door," Sam insists.

"Are you sure?"

"Yes. And I saw the wife and that son lurking around there."

"Interesting. I'd still be shocked if one of them nabbed it, bro. They don't seem like the type of people to do something like that."

"I'll just have to look through everything again tonight."

"Sounds good."

"You might have to call him again though," Sam casually mentions.

"Fuck, Sam. I don't want to talk to him. I'll just give you his number."

"I don't want him to have my number, Q."

"Well, fuck, bro, when did I become the babysitter?"

"Don't freak out. We will deal with it later," Sam says.

As we blaze down a road that leads outside Boston, my water bladder no longer is dispensing liquid.

"Oh, come on," I yell in frustration, yanking on the tube attached to the front of my backpack.

I come up on Sam waiting for me at a car repair shop. I stop and pull out the plastic water bladder from my backpack. The hose running from my mouth to the bladder isn't properly attaching at the base, so water is seeping out and falling into the bottom of my backpack.

"Told you I still got it," Sam remarks as I dry the water from the back of my shirt.

"What are you talking about?"

"My pace. I haven't lost a step."

"I can see you're never going to let that one go," I huff.

"Damn right!" he howls with a menacing laugh.

In 60 miles, Sam and I will be that much closer to the end of our journey. Having amassed over 1,700 miles already, we now have only three days and 130 miles left.

Around 2:00 PM, we stop for lunch at a sandwich shop near the New Hampshire border.

"Dude, look at your back tire," Sam exclaims while I finish off a grape iced tea.

"What do you mean?" I ask, following his finger to my back tire.

"That!" Sam wails.

When I turn around, I can see why Sam is so animated. My tire is decimated. White shreds are peeking through the black exterior.

"That doesn't look good," I observe, lifelessly noting another flaw in my apparatus.

"You mean you haven't noticed this before?" an incredulous Sam persists.

"Obviously not, bro."

"Well, what are you going to do about it?"

"I don't know, man. We don't have that many miles left. It doesn't make sense to buy another tire."

"You're just going to leave it like that?" Sam grimaces, as if I'm willing to spend the rest of the day wearing my shirt backward.

"I wouldn't quite say that."

"What does that even mean?"

"Sam, I'll figure something out. Let's just go."

Paranoia invades my brain. I don't know how to feel about this tire. Every bump in the road ahead suddenly feels more important. I don't want to hit even the smallest pebble for fear that my tire may then burst and send me tumbling all over the road.

We exit Massachusetts state lines still directly next to the Atlantic Ocean. Now we are entering New Hampshire. It's there we are greeted by a spectacular strip of waterfront stores that face toward the majestic ocean. Calming waves crash into rocks on the shoreline, and peace is temporarily found.

The roads along the strip are winding and narrow, but it's not like New York. No one is budging me out of the way so they can get to work. Cars accept my bike as part of the road by graciously moving over as far as they can. To my left, a group of kids are playing outside their parents' palatial home. The grass sways toward the ocean with tranquility.

Life hasn't felt this peaceful in weeks. This is how the trip should end; Sam and I able to take in the sights and not be concerned with anything other than embracing the short time we still have left.

We take another break at a gas station that has a Dunkin' Donuts attached to it. I go inside and order a sixty-four-ounce strawberry smoothie.

"This tastes so fucking good. I wish they had more of these Dunkin' Donuts in the Midwest," I tell Sam.

"You'd definitely get diabetes then," he jokes.

It's easier to appreciate everything today. I don't need to line up more hosts, and Sam isn't concerned with navigation. The signs will take us straight to Portland.

I'm waiting for a moment to appear, one that signifies the importance of this trip, a crescendo to the seven weeks of physical and emotional tumult we have endured. We didn't bike all the way out here just for the exercise.

"What are you thinking about?" Sam asks as we finish our drinks.

"I think you already know the answer to that question, dude."

"Amber?"

I neither nod nor shake my head.

"Q, you'll get over her. Just give it some more time," Sam advises.

The roads are busy when we arrive in Portsmouth around 5:00 PM, but the good news is our host Andy doesn't live on top of a hill.

"Did you guys have a nice ride today?" Andy asks when we pull up behind his building.

"It was good. We got to come in along the coast," I tell him.

"Oh yeah. That road is absolutely beautiful, don't you think?"

"It really is," Sam adds.

"Let's go to my office, guys. I'll show you around," Andy says.

He takes us upstairs and shows us the main room, a kitchen that has a dirty fridge, and the meeting room.

"You guys are going to have to sleep on the floor tonight," he mentions. "I know it's shitty, but my house is busy tonight."

This marks three straight days of sleeping in a tent or on the floor, which is comical because until now we have been lucky enough to sleep in beds or on couches.

"I have to get home," Andy says before heading to the door. "Call if you guys need anything, otherwise I'll see you in the morning."

I begin setting up a spot to sleep in the hallway that leads to the kitchen. My sleeping bag and spare clothes are going to be the only bridge between my skin and the slivers that will likely come from sleeping on a wooden floor.

Meanwhile, Sam's having no luck with his post-trip plans.

"Wow, I was not prepared for this," he says after his friend informs him that he can't stay with her while he looks for an apartment.

This comes on the heels of asking a not-so-gullible woman from Amtrak, "You mean, you don't have a closet or something I could just throw my bike in?"

Not being the most prepared guy in the world myself, I take solace knowing my flight home won't be filled with hiccups. It's easier to plan ahead, but Sam hasn't always done this, like the time he tried to convince an economics professor that, despite his academic shortcomings, he should still get a passing grade because he's a nice guy.

"Can't you just do me a little favor?" Sam asked after class that day as his professor stood in front of us and massaged his mustache.

The professor gave Sam that look, the one that says *come on, man. You can't be serious.*

Sam didn't pass that class.

CHAPTER 46

The next morning, I wake up at 4:00 AM to my back howling for mercy. The floor is so hard that falling back asleep is impossible, so I spend the next few hours listening to music and scrolling through Tinder until it's semi-appropriate to begin moving around the office.

Around seven thirty, Andy's coworkers arrive and begin logging in to their computers.

"Did you sleep on the bare floor?" one guy with gray hair and glasses asks me.

"Yes, unfortunately," I say, my tone echoing a polite hatred toward the man's inquiry.

"Where did you guys bike here from?" another man asks.

"Minnesota," I tell him.

"Wow, that's really impressive. I would have loved to do something like that if I didn't have a family," the man says.

"We're lucky. We just got done with school and figured this was the best time to do it before life gets too serious," I explain.

Sam comes out from the meeting room a short while later, and for the first time ever, he is ready to hit the road early. No morning bathroom sessions or prolonged conversation with a stranger. He is ready to bounce.

"You okay?" I ask to make sure he isn't insane in the membrane.

"I just want to get going," he claims, standing in front of the door.

"Oh my gosh, fellas," I begin, inviting ears into the drama. "Sam literally always takes his sweet time. I am actually shocked that today we are waiting on *me*."

"Geez, kid! Would you just shut up and get ready?" Sam jokes with a sheepish parting of his lips before disappearing down the steps and outside.

I begin packing up my bag and strapping my water bladder on my back.

"Andy was telling me last night that you wrote a book," one man says to me.

"I did. *The Long Road North*. It might be out right now actually."

"Well, let's take a look. Where can I find it?"

"Amazon or Barnes & Noble are going to carry it on their websites," I mention.

Sam coincidentally has returned from outside and joins in intently monitoring the screen.

"Is this it?" the man asks, pulling up a webpage that does indeed have my book available for purchase.

"That's my book," I softly mumble, the moment suddenly feeling monumental.

"I'm sorry," I slowly say, trapped in bliss. "I just didn't think I would ever do something like this."

The man adds the book to his virtual shopping cart. "There. Now you have one sale," he smiles.

"You totally didn't have to do that. I was going to send Andy a copy as a thank-you."

"Don't sweat it. Best of luck to you guys as you go through life," the jovial man smiles.

I walk downstairs in a cheery mood. It took me two years to write that damn book. I never sold more than a couple hundred copies, but it was a venture that led me down the path I currently walk. I didn't know if I would ever make it as a writer when the man bought my book. I just knew I had a chance.

Sam and I move over to a gas station to buy a few bottles of water and chocolate milk. He is still bitching at his friend when I walk out with a bag in my hand.

"I still can't wrap my head around this. All I need is a week," Sam continues to lament to his unforgiving friend. "Why can't you just do me a solid? I don't have a lot of money right now."

"Everything okay?" I ask after he hangs up.

"No. I'm going to have to call my brother or something. I don't have a spot to stay when I get back."

"You'll figure it out, Sam. You're a smart guy. But listen, a bike shop next to our building opens up in a few minutes. Let's go see what they can do with my tire," I tell him.

We walk around to the back of the store. Knowing my history with bike shop employees, my sanity will soon give way to anger and frustration. As I put my kickstand down, I notice an employee shuffling bikes around a large rack.

"How's it going this morning?" I ask him.

"Good," he says with no emotion.

"Are you the owner?" I ask.

"No. What can I help you with?"

Despite my prolonged explanation, the man is of no help and directs me to someone else.

"What kind of tire are you looking for?" a younger, muscular man in the repair section of the store wants to know.

"Just something to get me to Maine. It doesn't have to be nice, as long as it doesn't have holes in it," I joke.

The guy isn't much of a talker, but he quickly goes to work on the back tire. He stops to marvel at just how beat-up the busted rubber is.

"And how long would you say you rode with this?" he asks, holding up the tire in the way Sam dangles dirty socks.

"I really have no idea."

"Wow. This thing should have failed four hundred miles ago."

"That makes me feel lucky," I tell him.

The man finishes with my tire, and we walk over to the cash register. It's there that he throws me a bone.

"I'm not going to charge you for labor," he says.

"That's not necessary. I can pay," I tell him.

"It's fine. Don't worry about it," he assures.

"Okay. I really do appreciate your gesture."

"I know you do," he says, and then swipes my credit card.

I shouldn't reject charity. There is enough money in my bank account for a few Chipotle burrito bowls, and maybe a week's worth of groceries after I move into my new apartment. After that it's anyone's guess when stability will find my bank account.

I also have a massive debt in the pay-it-forward department. The free food and accommodations from Warmshowers hosts are obvious, but it's the kindness that will cost the most. I owe the cycling community a place to rest their head on their way through Minnesota, a meal, plus a trick or two I've picked up along the way. Someone besides me has to benefit from these seven weeks.

Sam and I leave the bike shop and cycle toward Gorham, Maine, ready to cross state lines for the final time. The humidity isn't too bad today, and the sun isn't burning my skin to a crisp. There is even a slight breeze that quenches my desire to chug water every few minutes.

At a fork in the road, Sam is again nowhere to be found. I go right but soon realize I should have gone left. The backtracking isn't long, and soon we are in the parking lot of a grocery store eating lunch.

Today the miles are ticking off so fast, I wish the road would throw a few more hills our way just so the journey lasts a little longer.

As Sam and I eat our respective high-calorie meals, there is a certain harmony to the way we speak to each other. Our conversation is all banter, but that's how it's supposed to be even though we both know the end is near.

"I don't care what you say. You fucked four women on this trip, so you better be happy," Sam chides.

"Yeah, and it would've been double that if it wasn't for you!" I sarcastically yell.

"God, kid! I've never seen someone as cocky as you. A woman touches you on the shoulder and you think she wants to bang you."

"That's because it's true. Don't be a hater, Sam."

"Hurry up and eat that sandwich because—"

"Because otherwise you're going to eat it," I laugh.

"No, douche. Because we're almost there. Only ten miles left."

"Sounds good. Just get us there. No more talking," I joke.

With just a couple miles to go, my brain takes a stroll down memory lane, back to when I did my first long-distance ride.

THE LONG ROAD EAST

It was 2014 and my buddy Rhino and I needed a new drinking spot, so we busted our asses through desolate northern Minnesota to arrive in a small town. When we got there, the first move was to stop at a bar, grab a burger, and then begin pounding Grey Goose Red Bulls.

Intoxication being the barometer, our night was a success. It's taken me until now to realize the real victory that night had nothing to do with alcohol.

The win was further developing a strong friendship with Rhino, a guy who in his own unique way inspired me to get out of my comfort zone and think big.

If it wasn't for his excessive nagging and cries for me to stop whining about my sore lumbar, I would have quit biking and retreated back to my comfort zone. Rhino brought out a part of me I didn't know I had, and then over the next year our ambitions brought us to other cities, the final and most grandiose being the Canadian metropolis of Winnipeg.

If I hadn't followed his lead, I might still be stuck unloading boxes at a shipping company and hoping something better smacked me in the face. Rhino, the man who was once my enemy, taught me to believe in myself and keep asking for more. It's a huge reason why I'm currently in Maine.

"We should have invited Rhino," I tell Sam when we come to a busy county highway.

"I don't understand. Why are you saying this now?"

"Good question. I just got to thinking about it. He probably would have said no, but we still should have asked."

"He never would have come."

"I know, but we still should have asked. He deserved that much."

We then shift our attention to the road ahead. After crossing the busy county highway, there is only one more turn before we reach the home of a Warmshowers host by the name of Lyle.

We come within ten feet of his driveway. The traffic blazes past on both sides with urgency. Rocks, pebbles, and old tar are bouncing around as an endless number of tires rub against the pavement. Waiting for all the cars to stop would take a lifetime.

"Fuck this," I agonizingly tell Sam. "Let's go up and then turn around."

We ride forward for a few more minutes before there is a gap in the traffic. Then we turn left and hook around so we can enter Lyle's driveway from the other side of the road.

As soon as we park our bikes in his driveway, the trip is all but over. There are fifteen miles left tomorrow, but they'll be easy. Today was the last haul, the last time we will have earned a pillow to lay our heads on.

"You want some kefir?" Lyle asks while he goes through the fridge. "It has probiotics."

"I'll take a glass, please," I say, and then after downing the thick liquid, Lyle pours me another round.

Lyle's girlfriend joins us for dinner later that night. Afterward we all sit in the living room and trade stories. Sam's epic rant in Michigan gets revisited, and of course I have to enrage Sam and throw in the part about Ruxin.

"You two seem to have a special chemistry," Lyle notes as Sam sets down his phone and begins admonishing me. "Make sure you guys appreciate each other."

My heart feels fuzzy. "A lot of blood, sweat, and hangovers to get to this point," I tell Lyle.

The night wouldn't be complete if Sam didn't have a beer. He flips the cap off and lets the pilsner stimulate senses his body has been conditioned to enjoy. Had Sam been an AA advisor and me a practicing Catholic, this trip would have sucked. He likes his beer and I like my women, and without these two extracurriculars things just wouldn't be the same.

One more ride, and then we are done.

CHAPTER 47

Everything today will be for the last time. The last time I wear bike shorts. The last time I pinch my tires to check the air pressure. I no longer will have to sit on my seat and wince while massaging my way into a comfortable position, my butt cheeks in agony while swiveling and rupturing one of the few zits that have formed above the back of my thigh. It's difficult to sit for seven weeks straight.

There are many other things that will be experienced for the final time. Not having a lunch break after twenty miles. No Sam barking at me to pick up the pace. No arriving at a house and having to already start thinking about the next day. All that now belongs to the past tense.

I send our last hosts a message to notify them of our impending arrival. It's not long before we pass a Domino's Pizza and are in the business district of Portland. The closer we get, the more I want the miles that are ticking off to come back.

Comic relief unfolds when we get turned around near a hospital, ending up near a basement parking lot.

"Sam, even on the last day, you still get us lost," I rib.

At the next stoplight, further reflection commences.

"You know, if you were to tell me four years ago that I would be standing at this stoplight with some guy who wears too much red, I would have never believed it," I tell him, harkening back to the day I first met Sam as he walked through the door of our Spanish class wearing all red attire.

Sam shakes his head and smiles, preparing for the light to turn green.

"But goddamn, Sam, this has been the best time of my life. Thank you for everything, brother," I finish.

"I still don't think I was wearing *all* red," Sam smirks.

The light changes, and we ride toward the final destination of summer 2017. It's not long before we are at another stoplight, but this one isn't filled with reminiscence.

It starts as routine as ever. Sam's hugging the left part of the turn lane. I'm behind him, more centered. A small hill is up ahead that we are fortunately going to avert by turning. There is a park on the left that sits still under the morning sun. Nothing about this moment seems significant. And that's when it happens.

A black pickup truck screeches around a right turn. I look left and am taken aback by the roar of the engine. The truck comes closer, not turning enough for my liking. It feels like slow motion, watching the vehicle with tinted windows move closer but still not veer away. I can't see a face behind the windshield, and the truck continues to barrel toward Sam.

He sees the same thing, but something bad is going to happen unless Sam makes a move.

Sam then whips his left leg forward, but by now there is no way he's getting off his bike in time. And the truck still keeps coming, like the person driving can't see what's in front of them. The truck crosses over the bike lane, then its own lane, finally penetrating our left turn lane.

Smack!

The car crashes into the back of Sam's bike, missing his leg by about two feet. The collision produces a swift clank, and a few chips from the back of Sam's bike fling in different directions. I look on in fear, even while seeing that no part of the truck connects with Sam's body. The truck keeps going, a *Mad Max*-type vehicle that isn't going to stop for anything.

"What the fuck!" Sam screams, looking back in the direction of the truck.

The truck must have heard his cry because the driver pulls over a couple hundred yards ahead, but no one immediately gets out from the vehicle.

"Are you okay?" I yell to Sam, my eyes bulging with fear.

He's not maimed, but the anger in his eyes makes me wonder if he'll ever feel whole again.

"Dude, are you okay?" I yell again.

Sam doesn't reply, and in my own discombobulation I head for the truck, ready to unleash a flurry of choice words. As I get closer, a military license plate comes into view, and then the sheer size of the vehicle strikes another nerve as I move to the passenger window. The whole time I'm running through what I should say, but when I get to the window, suddenly no words seem appropriate.

My eyes meet those of the driver. He's sitting in a wheelchair. All the insulting things I had wanted to say no longer seem fair. My mind can only guess at the unfortunate events that have led to this man being here.

"So, like, what happened, man?" I fumble.

"I didn't see him," the man says, sunglasses covering his eyes.

"You didn't see him?"

"I didn't see him," he repeats in the same listless tone.

He seems to be in as much disbelief as me. Sam comes over and approaches the window, and I walk toward him.

"Be careful," I advise, touching his arm ever so slightly in hopes that he catches my message.

Sam gets to the window, my warning no match for his heightened emotional state.

"What do you mean you didn't see me?" Sam fumes.

"I didn't see you," the man again repeats.

"We were in the middle of the turn lane!"

"I know. I didn't see you," the man reiterates, his tone still as emotionless as before.

"Well," Sam huffs, looking around like he has no other options. "You could at least say sorry or something."

The man doesn't apologize, instead hanging his head at the unbearable shame he must feel. A few city workers come over, and now a throng of people have formed in the park.

"We called the police," a worker with tattoos covering every inch of his arms alerts.

"What? Why?" Sam complains.

"We have a legal obligation," the worker explains.

Today wasn't supposed to be like this.

Soon a police car arrives, and Sam begins talking with an officer. My first thought is how we have been forever robbed of our epic denouement.

"Excuse me, Officer," I say, walking over and intruding on his paperwork as he sits in his car. "How do we go about getting my friend's bike fixed? Is that something the man's insurance fixes, or do we have to have a conversation with the guy who hit us?"

"Insurance will be notified, but you likely will have to deal with the man that hit you," the officer says.

It's then I realize that peace has to be made. After a quick and friendly conversation, the old man who hit us agrees to drive Sam to the bike shop.

"You can just ride your bike over there, right?" Sam asks me.

"That's not a problem," I say, but truthfully getting on a bike now has me scared shitless.

"Then I'll see you in a bit," Sam says, then he and the man drive off to the bike shop.

I fortunately make it through the next few miles unscathed, arriving at the shop to find the old man sitting outside next to his truck.

"Hey, how's it going?" I ask, getting off my bike and still towering over him.

"I'm Glenn," he says.

I bend over to reach his hand.

"It's nice to meet you, Glenn."

"Look, I'm really sorry about all this," he starts.

"Honestly, don't worry about it," I say. "We might have overreacted earlier. We were just juiced up because of everything that happened."

Glenn then starts to feel more comfortable. Disaster has been averted, and it looks like everything will be okay.

"I'm just glad Sam wasn't hurt," he says.

"Me too. And that's really the important thing. Everything concerning the bikes is replaceable."

"Should we go inside and check on him?" Glenn asks.

"Yeah, let's go."

Inside, one of the workers offers Sam bad news. "You're going to need a new rim," the employee says.

"Get the best one you can," Glenn tells Sam.

"Oh, I don't need the best one. Just something that works," Sam says.

"No," Glenn urges. "Get the best one."

The worker then tries to piece everything together. "How do you guys know each other?" he asks.

"I hit him with my truck," Glenn confesses, pointing at Sam.

"Oh," the man says, unsure of exactly what he's dealing with.

"It's a long story," I tell him.

I begin to think about what would have happened if the GPS hadn't turned us around at the hospital. This is a rabbit hole of dark thoughts, but we might not be here if Google hadn't erred, our trip over and Sam's bike mangled. Sam also would not have made a last-ditch maneuver to save his own leg, and I would not have witnessed him being milliseconds away from getting pulverized.

Then again, had none of this happened, it is unknown how fate would have showcased her evil. As my dad later said, "You ride a bike long enough and things like that are bound to happen."

My dad is right. Certainly, we didn't ask to be put in compromising positions on this trip, but our mere presence was enough to invite horrible things to happen.

"And you want to get your bike boxed up, correct?" another worker asks me.

"Yeah. I have a flight in two days."

"Sounds good. I just need you to sign a few things."

After taking care of business and leaving our precious bikes in the hands of the shop, the three of us exit the store. I then take out my phone.

"We'll be there soon," I text Rob, our Warmshowers host.

I should mention the circumstances surrounding our arrival but figure it's better to wait and discuss everything in person. This gives me more time to conjure up an explanation for why we will be arriving in a truck, devoid of our bikes.

After a few minutes of driving, it's evident Glenn should not be operating a vehicle.

"Oh, was that a stop sign?" he asks, braking once he's all the way into one intersection.

"I think it was," I cringe.

The front seat now feels awfully lonely, like I'm suddenly an accomplice to whatever catastrophe happens in the next few miles.

"We're not in a hurry, Glenn," I tell him, but he brushes off my sentiment like a taxi driver in rush hour.

We eventually turn onto Drew Road, my heart both leaping and sinking when I see how beautiful the ocean is. The view is amazing, although it's cheapened by entering inside this burly vessel.

We begin unloading equipment from the back when Rob walks out his front door with the exact look I hoped he wouldn't have.

"What's going on here?" he questions. "Are you guys traveling by car?"

My stomach curdles, quickly trying to think of the best way to describe the situation.

"It's a long story, Rob," I begin. "I figured it would be best to explain everything in person rather than via text."

"What do you mean?" he questions.

Having Glenn within earshot, I try to exercise a level of tact. "Basically, Rob, we were four miles away and then Sam got hit by a car," I explain. "His bike is in the shop and that's why we got a ride here."

Rob processes my explanation, despite many key components still missing from the story.

"Wait a second. First off, who hit you?" he asks.

I open my mouth but don't quite know how to respond.

"Actually, the guy in the car," I say, pointing at Glenn.

Rob's face twitches ever so slightly, and Glenn begins waving at us.

"I see," Rob says, the words slowly coming out of his mouth.

"It's a long story, and I'm happy to explain it all to you," I assure him.

"Yes, of course. In due time."

THE LONG ROAD EAST

Rob walks over to Glenn, and they talk inaudibly.

"This is definitely awkward," I tell Sam.

Rob soon comes back over. "I don't think you guys could have had worse luck," he says. "Come on. Let's go inside."

Glenn then turns his truck around and drives away.

I dreamed this ending would be filled with grandiosity, but no fireworks or champagne are waiting at the finish line. After I explain everything to Rob in greater detail, he soon continues with the rest of his day.

"If you want to wash your clothes, the machine is upstairs," Rob says.

"That would be awesome," I reply.

"You might want to air dry your clothes though," Rob says. "The dryer is old."

While the clothes wash, Sam and I go onto the upstairs deck and look out at the ocean. A construction crew is working on a house across the street. For seven weeks, I was able to feel special about being young and free. Now seeing the men at work is a harsh reminder that soon I will be joining the rest of society in the working class. I'm no longer a student with a life to go back to. I'm an adult who has to build a foundation.

Painfully, I go downstairs and clear out my belongings from Joad. My plan all along was to give her away at the end, so I walk her to the front of the yard and stick a free sign on her.

"I don't know if anyone is going to take it," Sam blasts when I walk back inside.

"Don't be a dick," I tell him.

Rob senses my disappointment. "I'm sure someone will take the trailer," he assures.

Joad's value can't be determined by appearance. I don't want to let her go, but there is no way she's coming back to Minnesota. Her departure is now a waiting game, and it hurts like hell.

CHAPTER 48

A morsel of my soul dies when I look out the window and notice Joad is gone. It must have happened during the night. This moment should be joyous, letting her go so she can bring happiness to another person. Yet when I set her outside yesterday, I didn't expect anyone to actually take her. I didn't think anyone would see this lump of elastic with a rusted frame as anything more than a chore that belonged in a dumpster.

They wouldn't know what Joad had been through, all the miles we spent together discovering more of this beautiful country. They wouldn't know she went almost a whole day with one bum wheel, persevering through the hills of Pennsylvania.

"I think a young family a couple houses back took it," Rob's wife Martha says. "They have kids, so it will be put to good use."

"That's good, I suppose."

The day is fairly slow. Besides me raiding every section of the fridge and cupboards, not much happens.

"Do you want to go see this new shark movie?" I ask Sam.

"You and shark movies," he laughs before formally declining my verbal invitation.

After his rejection, I go upstairs and sit on the deck overlooking the ocean. It's a reminder of how far I have come; not only on this trip, but in life.

When I think back on these seven weeks, everything happened so fast. It was like being on a rollercoaster that never stopped. But even when you're on a rollercoaster, you can still catch a few glimpses of something worth remembering.

THE LONG ROAD EAST

In no particular order, I remember a guy named Buddy in NYC who said he couldn't host us because he was spending time with his parents.

I remember the smell of the room in Tarrytown after having sex.

I will never forget the way the rain fell off my face all throughout Michigan, or the reluctance in Sam's eyes as I emotionally vomited at him just outside New York City.

I remember leaving a swimming suit in Connecticut, the same ones I thought perfectly accentuated my legs.

I think of that stupid dog in Wisconsin who bit Sam's foot, and wanting to go back and punch it.

It's hard to shake the visual of seeing the look in my own eyes when having sex in Milwaukee.

Everyone in the Midwest encouraged Sam and I to keep going, but people out East usually only offered physical support.

I remember a guy in NYC asking us where we came from, and then exiting my sight moments later.

And a truck transporting alcohol in Wisconsin that struggled up a bluff as much as I did.

The look on Sam's face that first night we slept in a trailer, clearly displeased with my riding.

The turmoil I felt in Pennsylvania, sickened with the way I treated my ex, and a host from the same state who asked God to forgive him for the sins of his youth while he was on his deathbed. I'm now hoping to be granted that same luxury.

I also think about not being able to grasp who I'm supposed to be, the man I am meant to become. I try to search for explanations to various feelings, but they are elusive. It's as if not enough time has passed and life has to be lived first, the best advice finally revealed when the universe feels I'm ready for it.

But most importantly, when I think of these seven weeks, and all the events, I think about where Sam was when they happened. I think about the disappointment I caused him, and the guilt when thinking of ways his journey could have been made easier.

He's my best friend. The best way I can put it is I love the guy. I loved him for putting up with me, for not giving up when it was clear from the start we would be riding at different speeds; for telling me things I never wanted but needed to hear; and most importantly, for simply being there, serving as a partner for all the great times we shared. I'm sure one day I'll call him, we'll catch a drink, and simply laugh as we relive the best seven weeks of my life.

CHAPTER 49

Seven weeks earlier, I didn't want to get on my bike. I only wanted to pick up Sam and bring him back to Osseo, Minnesota. Now that I'm a plane ride away from returning to where it all started, it's hard to accept that I won't *have* to jump on my bike ever again.

All my bags are packed, Joad is someone else's responsibility, Sam has a train to catch, and everything invested into this trip is coming to an abrupt end.

My stomach is in knots as I walk downstairs and attempt to eat a bowl of Raisin Bran. I can't take more than a few bites because the cereal is sticking to my throat. I guess heartbreak has that effect.

Over a year of planning. Multiple fights with Sam about logistics, money, women, and equipment. None of that matters anymore. In a few minutes, Sam is about to walk out the back door, beginning a tenuous future for our friendship.

"We should get going. We won't want to miss your train," Martha tells Sam as the seconds hand on a clock above a bookshelf inches closer to eight thirty.

And then my emotions become unhinged. I'm looking directly at the soppy mess in my cereal bowl, trying desperately to hold back tears that just won't die. I'm not sad because the journey is over; it's that deep down, in the smallest minutiae of buried realities, I know that this trip has to end.

A tear falls off my eyelid into my bowl of cereal, and soon another, and then another. I can't hold them back any longer. This 2 percent is about to turn into skim milk.

Until now, it hasn't been clear how integral Sam was to my happiness. He was always important and a guy I knew would stand by me through anything, but he's even more important than that, to the extent I don't know how to function right now.

All my trust, happiness, worries, and survival were thrust into his hands without asking for permission. Our partnership never should have worked, two kids from different backgrounds with polar opposite sex drives, yet here we are, in Portland, Maine.

I look his direction. I want to say thank you for bringing out the best and the worst in me, for showing how special and precious life is. It all sounds so good in my head, but then I choke. Tears play a part, but really, I have lost all artistic control. I can't express feelings without stumbling.

One of my last visuals of him is masked by the reality of knowing there will be no next year. No more drunken nights screaming at each other and playing NHL. No more taking trips together in pursuit of memories. No more playing the role of little brother. We are going to begin postcollege narratives, two separate lives filled with responsibility and hardships.

"I'll see you later," I whimper as Sam sets his bags down.

I give Sam a longer-than-normal hug, trying so hard to keep my composure, but tears are creating a mess I dare not try to wipe away.

He sees my pain and thankfully does not comment. Words come out of his mouth, but they are inaudible. My head is a clusterfuck of nausea.

When he closes the door behind him, I'm finally able to release all the tears that for so long left a burning sensation in my eyes and throat. Composure still has to be maintained, so I try my best not to make noise while Rob sorts through uncashed checks.

A million thoughts race through my mind, but one in particular stands out: *if I'm not doing things in life meaningful enough to make me cry, I'm not doing life right.*

An hour later Rob and I get in his car and drive to the airport. I check in at the front desk and then Rob accompanies me up the escalator. Feelings of loneliness and uncertainty continue to resurface.

"Well, this is where we part ways," I announce when we get to security.

Rob stops and smiles. "You make me laugh, Q."

"Why is that?" I ask.

He hugs me like a father and then hands over the rest of my bags. "It's just the way you look right now."

"What do you mean?"

"Sad, Q. You look sad."

I put my head down, embarrassed.

"Don't be sad," he says.

I don't respond, despite his best intentions.

"Q, look at me."

I peer into Rob's eyes one more time.

"Your life is just beginning," he tells me, and after a few seconds something magical erupts in my lungs. Spirits that have been burdened by hardship the last few weeks are awoken, and a smile forms on my face.

"That's funny," I say, picking up my tattered luggage and preparing to return to the life I have always known.

"What's funny? That your life is just beginning?" a confused Rob asks.

I nod and take one last look at Rob before heading for my plane.

"Those words you just said, about how my life is just beginning," I start.

"What about them?" Rob asks.

"I wrote them in a book once."

ACKNOWLEDGEMENT

Huge thank you to Melanie Bishop for working through the first draft of this book. She brought maturity to a person and story that was severely lacking in that department.

This book also doesn't get published if not for the keen insights offered by Casey Nordbak. As one of my best friends, Casey is one of the most sophisticated and knowledgeable people I have ever had the privilege of knowing. He's not only challenged me to become a better writer, but also a better member of society.

Lastly, thank you to my friend Tessa Carter. This book doesn't get published without her care and guidance, a reality that I am forever grateful for.

ABOUT THE AUTHOR

Quentin Super is an author, ghostwriter, and blogger.
 His first novel, *The Long Road North*, debuted in summer 2017.

In conjunction to his personal work, Super now professionally ghostwrites for clients across the globe who are seeking to share their respective stories. All ghostwriting inquiries can be emailed to Super via his website (www.quentinsuper.com).

Super is currently living in Minnesota as he works on his third novel, this one about his yearlong adventure in Beijing, China.